James Wharton McLaughlin

Fermentation, Infection, and Immunity

A New Theory

James Wharton McLaughlin

Fermentation, Infection, and Immunity
A New Theory

ISBN/EAN: 9783742817792

Manufactured in Europe, USA, Canada, Australia, Japa

Cover: Foto ©berggeist007 / pixelio.de

Manufactured and distributed by brebook publishing software
(www.brebook.com)

James Wharton McLaughlin

Fermentation, Infection, and Immunity

FERMENTATION,

NFECTION AND IMMUNIT

A NEW THEORY OF THESE PROCESSES,

*hich Unifies their Primary Causation and Places the Expla
tion of their Phenomena in Chemistry, Biology, and
the Dynamics of Molecular Physics.*

BY

J. W. MCLAUGHLIN, M. D.,

AUSTIN, TEXAS.

COPYRIGHT, 1892.

AUSTIN, TEXAS:
EUGENE VON BOECKMANN, PRINTER AND BOOKBINDER.
1892.

PREFACE.

Notwithstanding the fact that much valuable information has been acquired concerning fermentation, infection and immunity, and various theories have been advanced at different times which explain, more or less perfectly, large groups of phenomena of these processes, it must, nevertheless, be admitted that these theories are all seriously defective. They not only fail to give a philosophical explanation of the phenomena involved, but are unable to account for that intimate relationship which common observation, and that based on scientific investigation, has long recognized to exist between the processes in question. The obscurity which surrounds the intimate nature and causation of the phenomena of these important subjects, and the practical and scientific value of their exposition to medicine, and, especially, a belief that the new theory which we submit offers a rational solution of these problems, have induced me to offer this volume to the public.

The theory of fermentation, infection and immunization, which is herein elaborated is called, for the want of a better name, the physical theory, for the reason that the principles which underlie and determine these processes are believed to be mainly physical; those principles, or laws of motion in matter, which have enabled us to correctly interpret the mysterious readings of the spectroscope, and which, also, led up to the discovery of the telephone and phonograph; are the same principles of science—the same laws of molecular dynamics which, I contend, are mainly instrumental in the production of the phenomena referred to. They, however, are not the only agencies concerned in the causation of these processes—chemistry and biology are equally important.

The aim and purpose of this essay, then, is to show that the accepted principles of molecular physics, and those of chemistry and biology, if supplemented by legitimate deductions from them, are amply sufficient to account for all the known phenomena of these processes, and also, to explain their relationship and intimate nature.

My first effort to place the causation of these processes on a physical basis was made in 1887, in an article entitled, "The Etiology of Acute Infectious Diseases," which appeared in *Daniel's Texas Medical Journal*, the same year. From observing how nature accomplishes physical results by physical laws in other departments of science, I became convinced that chemical and physiological action occurring within animal and vegetable structures offers no exception to this law of nature, and, that a solution of the problem involved in the causation of fermentation, infection and immunization could not be arrived at except by investigating these subjects from a physical standpoint. Pursuing this line of thought, I soon ascertained by reasoning inductively from the molecular constitution of matter, and the accepted views of science relating to atomic and molecular vibrations in unvarying and distinctive periods of time, that many important phenomena of fermentation and infection are directly traceable to these physical laws. But, to my surprise, other and equally important phenomena of fermentation and all those of immunity, refused to fall in line with the former, and yield their secrets to the physical method. It was at this stage of the work, while the problem was still surrounded by great difficulties, that I published the article referred to. Some time after this, while casting about for the cause of my failure to bring all these phenomena within the domain of physical law, it suddenly occurred to me that the effect of wave interference, that beautiful law, which Sir John Herschel says "has hardly its equal for beauty, simplicity and extent of application in the whole circle of science," had not been considered in my calculations. Francis Galton says, "Few intellectual pleasures are more keen than those enjoyed by a person who, while he is occupied in some special inquiry, suddenly perceives that it admits of a wide generalization, and that his results hold good in previously unsuspected directions." This statement proved eminently true in this instance. When the effect of "interference" of ether waves which are produced by molecular and atomic motions, is considered in connection with the molecular structure of matter, all the phenomena of fermentation and many of infection can easily be shown to result from these agencies. And, when the evolution of organic forms from primordial protoplasm endowed with forms of energy and potentialities (a result of molecular and organic structure), is primarily based on the principles of molecular dynamics referred to, then the laws of inheritance will supply whatever evidence may be wanting to fully account for the causation of fermentation, infection and immunization, and the intimate relationship existing between them.

. The subject, in a more matured form, was next presented in a paper, in 1890, which I read at the Texas State Medical Association. The paper was entitled "An Explanation of the Phenomena of Immunity and Contagion, Based Upon the Action of Physical and Biological Laws." But I found it impossible to intelligibly and fully include a subject so complex and novel within the compass of a society essay, consequently, the paper was seriously crippled by its brevity. Notwithstanding this defect, it received from some of the leading medical journals of this country very favorable, if not flattering notices, and the complete article was translated into a foreign language and published in a foreign medical journal. The encouragement I received from such favorable notice induced me to be more fully elaborate and again publish an article on this subject. This I did during the last year, in serial numbers of the *Texas Sanitarian*. The present volume is made up principally of these articles, which have been somewhat corrected, and to which has been added a considerable amount of important new matter bearing upon this subject.

AUSTIN, TEXAS, November 15th, 1892.

CONTENTS.

CHAPTER I.

CHAPTER II.

CHAPTER III.

CHAPTER IV.

CHAPTER V.

CHAPTER V—CONTINUED.

CHAPTER VI.

CHAPTER VII.

CHAPTER VIII.

FERMENTATION, INFECTION AND IMMUNITY.

CHAPTER I.

FERMENTATION, PUTREFACTION AND INFECTION, THEIR
RELATION TO EACH OTHER, AND THE SIMILARITY OF
THEIR CAUSATION AND PHENOMENA.

There is, perhaps, no subject of greater scientific value to
the physician, the solution of which would insure greater
practical results for the good of mankind than that involved
in the phenomena and nature of contagion, and the phenomena and causes of immunity.

Contagion is the infective principle of that large class of
diseases called contagious—communicated, or in common
parlance, "catching." Immunity signifies that condition
of the body which opposes the development of contagious
processes; that enables the individual to successfully resist
the invasion of contagia; hence, a knowledge of the nature
of infection and immunity is necessary to a knowledge of,
and the means of controlling or preventing infectious diseases.

The phenomena of infection are, in fact, the laws of infectious diseases, and can be best considered in a study of
these diseases. They are as follows:

Infectious diseases, as the name implies, can be transmitted or conveyed from the sick to the well, from individual
to individual, from neighborhood to neighborhood, or from

country to country, by contact, perhaps by currents of air, by infected goods, merchandise, food, etc., by sail or steamships, railroad cars, and in other ways.

The infective principle is a portable and particulate substance; it may be destroyed by overheating—in some cases by freezing, and by that class of drugs called antiseptics or germicides. It is variable in its degree of malignancy; for example: during one season, the disease of which it is the etiological factor, will prevail, perhaps, as a mild epidemic; the next season as a malignant epidemic; hence, it can be modified by environmental agencies.

Infectious diseases have periods of incubation of one or several days, dating from the day of invasion, to that when the disease becomes manifest.

Infectious diseases produce only their own kind, and do this as unerringly as do animals or vegetables in their increase by generation.

The infective principle, that to which the disease owes its specific nature, is always largely increased in amount, several thousand fold, during the course of the disease; this is well illustrated in small-pox.

That class of diseases known as the acute infectious are self-limited in their duration; their course is typical; if the patient can withstand the infection the disease will "run its course" and be terminated by the action of its own laws, provided intercurrent diseases, which are not necessary results, do not turn the scales against the patient.

One attack, in many of these diseases, will render the individual immune from future attacks of the same disease; or immunity may be secured by other methods which will be discussed further on.

This group of remarkable characteristics has caused infectious diseases to be classed by themselves, and they have excited, even from early times, much speculation as to the nature of the contagium or contagia which cause them.

There are no chemical bodies, gases or miasmata which can increase themselves by self-multiplication and growth as contagia do; hence, it is safe to say that they are not chemical substances.

Now, these propositions are so clearly in harmony with the laws of reproduction and transmission of species as commonly observed, that their truth appears almost self-evident. That they will be assented to by the students of natural science, goes without question; but it is within the present generation, that they were vexed questions which were involved in that of spontaneous generation, which, it is to be hoped, has been finally settled in the negative by the experimental investigations of Schwann, Schultz, Cagniard-Latour, Pasteur, Tyndal and others.

The theories of contagion which have received most attention are:

(1)   The catalytic theory of Berzelius;

(2)   The bioplastic theory of Lionel Beal;

(3)   The physical theory of Justus Liebig; and

(4)   The germ or microbe theory of the present time.

Unfortunately, the first—that of Berzelius—explains neither the one nor the other process. Catalysis—to which he ascribes the cause of the phenomena of contagion—remained as great a mystery as did contagion.

The bioplastic theory of Lionel Beal assumes the existence in the blood and tissues, of microscopic particles of formed material which he terms "bioplasts;" and claims for them power of self-production. These bioplasts, it is taught, are subject to abnormal changes, and when in this condition, are the cause of infectious diseases;—an ingenious theory, but founded upon neither known facts nor analogies, and inadequate to explain the phenomena of contagion; hence, it has never been received with much favor.

The theory of Justus Liebig claims that, when proteid substances undergo retrogressive metamorphoses, their constituent elements—molecules—will impart to other substances the same motions or activities which they themselves are undergoing; thus tissue elements, for example, in small-pox, will impart to other tissue elements in a healthy condition, the same changes which they are undergoing, and thus produce small-pox.

This theory fails to explain why disease, when once established, does not invariably end in the death of the patient; or why one attack often gives the individual immunity from other attacks of the same disease.

The germ theory assumes that all infectious diseases are caused by microbes or bacteria. These are one-celled microscopic organisms which belong to the vegetable kingdom and are in fact regarded as the contagia of contagious diseases, and which differ among themselves, as do the diseases which they cause; this is the theory of the present day, and of all others, the one best able to explain the phenomena of contagion, and is best supported by demonstrable facts.

The casual relationship which a given bacterium bears to a given disease is determined by a strict observance of the following three rules of Koch:

"*First*,—It must be proved to be present in all cases of the disease in question.

"*Second*,—It must further be present in this disease, and in no other; since otherwise it could not produce a special definite action.

"*Third*,—A specific micro-organism must occur in such quantities, and be so distributed within the tissues that all the symptoms of the disease may be clearly attributed to it."

"When a bacterium is found to conform to these three rules, we may fairly conclude that it stands in a particular intimate relation to this disease, and the probability of its being also the cause of this disease, is so strong that it approaches very near to certainty.

"The last link in the chain of evidence is, of course, supplied only by a successful transmission, before the overwhelming force of which all opposition must yield."*

The bacterium found to be present in any given disease, and supposed to be the cause of that disease, must be isolated before its biologic habits and pathogenic properties can be properly studied. To do this is not a simple matter, for it must be grown in artificial media that has been thoroughly sterilized, and under precautions that will prevent contamination with other bacteria, in order that a pure culture of the organism may be obtained; the form of the bacterium, its manner of grouping, its methods of reproduction, its color, its behavior to food media, its products, and finally, the result of inoculating it into susceptible animals must be determined.

Now, there is a large number of these bacteria which are known to be pathogenic. They have all been carefully studied; they conform to the three rules of Koch, and in some cases, inoculation of the bacterium has always produced the disease of which it is the undoubted cause. The following named pathogenic bacteria are taken from Frankel's Bacteriology, to-wit: "Anthrax Bacillus, Bacillus of Malignant Œdema, Tubercle Bacillus, Lepra Bacillus, Syphilis

---

(*) Text Book of Bacteriology by Carl Frankel, M. D. 3d American Edition, page 152, et seq.

Bacillus, Bacillus of Glanders, Asiatic Cholera Bacillus, Finkler-Prior's Vibrio, Vibrio Metschnikoff, Emmerich's Bacillus, Bacillus Typhosis, Spirillum of Relapsing Fever, Plasmoderm Malariæ, Friedlander's Pneumococcus, Frankel's Pneumococcus, Diphtheria Bacillus, Bacillus of Rhinosclerma, Pyogenic Bacteria, Staphylococcus Pyogenese Aureus, Staphylococcus Pyogenese Citreus, Streptococcus Pyogenese Bacillus Pyocyaneus, Bacillus B (Ernst), Gonococcus, Tetanus Bacillus, Bacteria of Septicæmia Hemorrhagica, Bacillus of Hog Erysipelas, Mice Septicæmia Bacillus, Micrococcus Tetragenous."*

A close analogy in many particulars is found to exist between the process of fermentation and putrefaction, which is a variety or form of fermentation, and that of infection. Both are caused by one-celled micro organisms.   However, long before this fact was known these processes were thought to be, in some way, closely related.   A review of the history and phenomena of fermentation will not only bring into prominence the close relationship, if not identity of cause existing between this process and that of contagion, but will also enable us to understand some of the phenomena of

---

*To this list I must add the Micrococcus of Dengue Fever. During the summer and fall of 1885 an extensive epidemic of dengue fever prevailed in this and other Southern States.   As the disease presented many characteristics of microbic origin, I was led to make certain bacteriologic investigations, which resulted in the discovery of a micrococcus in the blood of persons suffering with this fever.   It is true that my work has not been verified by others, but it is also true that it has never been disproved; in fact, no opportunity has offered, since that time, to do this work, as that was the last epidemic of this fever that has visited this State.   I believe that the micrococcus of dengue fever is pathogenic and the cause of this disease for the following reasons:

(1.)   It was found in the blood of those suffering from the fever

the latter, when the same phenomena occur in the much more familiar, if not better understood, process of fermentation. Fermentation is a chemical term which, in accordance with its derivation,—*fervere* (to boil)—was originally applied indiscriminately to all chemical changes involving the effervesence of a liquid. In its modern application it is used to denote the changes which certain complex organic materials are made to undergo by the action of peculiar bodies called "ferments." These are of two kinds : organized, or living, and unorganized, or non-living. The character of fermentation that will occur in any given case will be determined by the nature of the substance acted upon, and by the nature of the ferment; thus vinous fermentation is caused by the action of yeast cells upon grape juice, or other sacchariferous fluids. The bacillus *acidi lactici* converts the sugar of milk into lactic acid. (This is what occurs whsn milk turns sour.) The *micrococcus aceti* converts watery solutions of alcohol into acetic acid or vinegar; this sometimes occurs in the souring of wines; the various methods of making vinegar are based upon the power of this microbe to convert alcohol into acetic acid. The *micrococcus nitrificans* is the bacterium that converts the ammonia

whom I examined (about forty). (2.) A pure culture of the micrococcus was grown upon artificial food media, and its morphological characteristics, which differ from anything heretofore discovered, were observed. (3.) The organism was found in the blood in sufficient quantities to account for the disease. (4.) This organism has never been found in any other disease.

I feel authorized in making this statement from the fact that no other known micro-organism groups itself in the same way as the dengue micrococcus. The different group-forms of this bacterium which have been photographed from the same slide, i. e., group-forms of one and the same variety of coccus, would indicate that this bacterium is pleomorphic in its biologic habit.

compounds found in the soil, or placed there for fertilizing purposes, into saltpetre, which is appropriated by growing vegetation. The *bacillus amylo-bacter* is the the organism that causes butyric acid fermentation. In addition to the important part it plays in the manufacture of cheese, it is an especially active agent in destroying the cellulose of cell membranes in the decomposition of decaying plants. In the process of rotting hemp, flax, and other textile plants, in order to obtain their fibre, this microbe plays the all-important part. Many other examples might be given to prove that the kind of fermentation is largely determined by the nature of the ferment, and that these apparently simple microscopic cells, which constitute the ferments, differ greatly in their seemingly homogeneous structure,—evidenced by the difference in the fermentations which they cause; but as this subject will require further notice later on, it will be passed for the present.

The accepted theories of fermentation, and the history of their evolution, will be most easily described by giving the reader the history of vinous fermentation, which is the most familiar and best understood of the fermentive processes.

Grape juice, when perfectly fresh, is a sweet yellowish liquid which is made perfectly transparent by filtration through bibulous paper. In this condition, if excluded from the air, it will remain, unchanged, for an indefinite length of time; if, however, a portion of the unfiltered juice is added, or if it has not been filtered, the fluid will soon become turbid; first an active commotion in the liquid will take place, carbonic acid gas will be evolved, and the temperature of the fermenting liquid will rise above that of the surrounding air. After a time the effervescence will cease, the solid matter which has been suspended in the fluid, and which gave it the turbid appearance, will sink to the bottom as a slimy mass, and a remarkable change in the quality of the juice will have occurred;—its intensely sweetish taste will have disappeared, and it will have acquired that of wine. The sugar which the grape juice first contained will be greatly lessened, if not totally destroyed, and a new substance,—one that the juice did not at first contain—alcohol, will have formed.

A remarkable increase of the solid parts of the fluid has also taken place. This substance (called yeast) is largely composed of microscopic vegetable cells called saccharomyces cervisiæ. At the beginning of fermentation it is present in a scarcely perceptible amount, but it rapidly increases as the fermentation progresses, and is finally deposited as a slimy mass when the fermentation is arrested.

In the disappearance of the sugar and its replacement by alcohol, it would seem, even to the ordinary observer, that the former had been converted into the latter, inasmuch as chemical reactions are simply the rearrangements of the ultimate atoms and molecules of the bodies concerned;—which atoms and molecules are assumed to be uncreatable and indestructible. That the alcohol and carbonic acid (dioxide) are actually formed from the constituents of the sugar, and together, represent the amount of sugar which has been destroyed during the fermentation, becomes demonstrated when we examine the equation representing this reaction; for example:

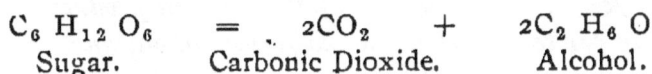

$$C_6 H_{12} O_6 = 2CO_2 + 2C_2 H_6 O$$

Sugar.　　　　Carbonic Dioxide.　　　Alcohol.

The reaction consists in a disruption of the sugar molecules and their rearrangement to form alcohol and carbonic acid, yet it will be observed that the yeast-cells do not figure in this equation, although they perform a very important part in the process; in fact, the fermentation cannot take place without them; neither a solution of sugar nor sterilized grape juice can be made to undergo the chemical reactions above described, by any known chemical agency.

Milk, if left to itself, as everybody knows will soon turn sour; the sugar of milk, which it contains, is changed into lactic acid through the agency of a microbe,—a one-celled plant called *bacillus acidi lactici*. The reaction that occurs is represented by the following equation:

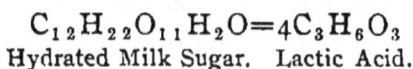

$$C_{12}H_{22}O_{11}H_2O = 4C_3H_6O_3$$
Hydrated Milk Sugar.   Lactic Acid.

The same elements in the same proportions are contained in both substances,—milk, sugar and lactic acid. In fact, as demonstrated by the equation, the sugar is decomposed, its molecules liberated, and their rearrangement forms lactic acid. This chemical reaction, like the previously described one, cannot take place except when caused by its ferment, the *bacillus acidi lactici*, and yet this ferment, like that which causes vinous fermentation, does not figure in the chemical equation. These are not exceptional cases; they serve only to illustrate a law that *ferments excite chemical change in other bodies, without themselves undergoing chemical change.*

There is another law of fermentation of great importance, which is as follows: *All ferment products inhibit their respective ferments and an accumulation of these products beyond a certain per cent. will arrest the fermentation, and this can not be re-established until the per cent. of accumulation is reduced.* Thus, the alcohol produced in the vinous fermentation brings this process to an end, and it cannot be re-established until the alcohol or a large portion of it has been removed from the fermenting fluid. If the grape juice contains a large proportion of sugar, its fermentation may become arrested by the resulting alcohol, before all the sugar has been decomposed. When this happens, as it often does, when the sweet Southern grapes are used, the wine will be of the sweet variety. On the contrary, when all the sugar is destroyed, the resulting wine will be sour. This, as before stated, is a general law of very wide, if not of universal application, the truth of which has been demonstrated time and time again. In the same manner that alcohol arrests the vinous fermentation, lactic acid will arrest the lactic acid fermentation; acetic acid will arrest the acetic acid fermentation; butyric acid will arrest the butyric acid fermentation; and the products of pathogenic bacteria, known as ptomaines, etc., will arrest the infections of which they are the respective causes.

One of the characteristics of infectious diseases, you will remember, is that they are self-limited in their duration. In the opinion of the writer this characteristic depends upon two causes, one of which is the law to which reference has just been made; the other will be discussed later on.

The study of fermentation by exact or scientific methods became possible for the first time when Lavoisier introduced the balance as an indispensible means of making quantitative chemical analyses. Prior to the use of the balance by Lavoisier, such substances as air, and all gases were regarded as imponderable, or unweighable matter, and their presence was not noted in the chemical analyses of that time. After its introduction and use it became possible, for the first time, for a chemical reaction to be quantitative as well as qualitative, and to take the rigid form of an algebraic equation. Long before this time, however, observers had noticed the outward forms of fermentation, and a few had speculated upon the hidden causes of this process. Notably among the latter was Stahl, the founder of the "phlogistin theory," who propounded a mechanical theory of fermentation which, considering the lack of scientific knowledge concerning this subject at that early time, must be considered a most remarkable production; it is substantially the theory afterwards advocated so ably by Liebig, and I believe it will furnish the basis for the theory of the future.

Stahl recognized the relationship which exists between the causes of fermentation and putrefaction, and ascribed both of these processes to a disturbance of the molecules in the fermenting body, which disturbances, he claimed, were caused by pre-existing molecular disturbances. After sleeping for upward of two hundred years, this theory was revived, and dominated the scientific world, under the leadership of Justus Liebig, until it was finally displaced, about 1867, by the germ theory, chiefly through the labors of Pasteur.

In 1680, Lieuwenhock, the father of scientific microscopy and the discoverer of the capillary circulation of the blood, discovered that yeast was composed of ovoid or spherical bodies, about $\frac{1}{100}$ millimeter in size. This, and other remarkable discoveries, he made with a simple lens which he manufactured.

About 1838, Schwann and Cagniard-Latour, independently of each other, with better microscopes than those used by Lieuwenhock, found that the bodies described by the Dutch scientist were one-celled vegetable organisms which multiplied themselves, and increased, in accordance with biologic law. From this, Schwann was led to believe that in some way the growth of the yeast cells was causative of fermentation.

In order to test this opinion by experimental proof, and to solve the problem of spontaneous generation, a subject in which he was deeply interested, he filled vessels with an in-fusion of flesh, which he knew would speedily swarm with microscopic organisms if left to itself. The flasks were then immersed in boiling water and kept there for some time in order to destroy all living organisms, or their germs, which the infusions contained. They were then hermetically seal-ed, before removing them from the boiling water, and set aside for observation. The contents of all the flasks which had been thus treated, remained absolutely sterile; when, however, air was admitted into any of them, putrefaction ensued and the infusion was found to contain innumerable micro-organisms; hence, he concluded that the air, or some-thing which it contained, was the cause of the putrefactive changes.

He next used flasks connected with long glass tubes. Into these flasks he placed his meat infusions which he sterilized by boiling; as before, the tubes were left open, but were kept constantly heated by the flame of a lamp; thus the air had ready access to the infusions, but, in passing through the heated tubes, all organisms which it contained were destroyed by the heat. The infusions subjected to these conditions always remained sweet and pure.

He next subjected vegetable infusions and other ferment-able liquors to the same treatment, and obtained the same results; hence, he concluded that it was not the air, but micro-organisms which it contained, that caused the fer-mentation.*

Schwann then prepared his infusions, to which he added certain drugs, such as corrosive sublimate, etc. (called anti-

---

*This method had, in fact, though empirically, been practiced on a large scale as a matter of industry, long before this time, in the pre-serving of fruits and meats by canning.

septics), which were known to be poisonous to infusoria and vegetable micro-organisms;—with the result of rendering the infusions unfermentable. Subsequent observers not only confirmed the work of Schwann, but added to it in important respects; for example: Helmholtz showed that oxygen evolved by electrolysis from water does not, like air, induce putrefaction; and what is more important, he showed that sterilized grape juice when tied up in a bladder, did not ferment when immersed in fermenting juice. An experiment of a similar nature, by Mitscherlich, proved that a septum of bibulous paper separating the two halves of a vessel containing grape juice, limited the fermentation to that side in which the yeast was placed. Some years later Hoffman accomplished the same results by a different method. He separated grape juice contained in a test tube into an upper and lower half by means of a cotton plug, and showed that fermentation was confined to that half in which yeast was placed.

It would be supposed that the evidence offered by these admirable and carefully conducted experiments would have finally settled the question of the causative relationship of micro-organisms to fermentation,—but such was not the case; they made but slight impression upon the scientific opinions of the day. In fact Liebig, whose theory was almost everywhere accepted, and if it had a rival it was rather the catalytic theory of Berzelius than the germ theory, ridiculed the idea that yeast was in any way necessary to fermentation. He clearly recognized that putrefaction and fermentation were analagous processes, but believed that both were caused by instability of complex organic molecules, such as albuminous compounds, which were constantly breaking down, and rearranging themselves into simpler and more stable compounds. This molecular disturbance, he claimed, imparted a similar molecular disturbance to other unstable compounds with which they came in contact, and thus gave rise to fermentation and putrefaction. Yeast, he claimed, was not directly concerned in the operation; if it assisted at all, it did so only in furnishing, from its dead cells, an albuminous substance undergoing a retrogressive metamorphosis; he ridiculed the idea that it was a living, growing ferment, capable of causing fermentation.

This was the status of the question when Pasteur, about 1857, announced the result of his labors in this field. He not only repeated and verified the experimental work of Schwann and other predecessors, but he modified, varied and perfected this work by a series of the most brilliant and convincing experimental proofs which successfully met all former objections, and forced the attention of all workers in this field of science. It was well known that wine, beer and other fermented liquors were liable to undergo other fermentations; thus wine often soured from acetic fermentation, or became thick and ropy from the viscous fermentation; in like manner beer often underwent destructive fermentations. These apparently abnormal phenomena of fermentation were thoroughly investigated by Pasteur. He found that abnormalities in fermentation, such as have been mentioned, are caused by the presence of other ferment organisms, which would remain dormant until conditions favorable to their activity were developed. In some cases several of these varieties of ferment organisms were found to be present in the same fluid, and would successively excite their characteristic fermentations as the conditions of the fermentable fluid changed. He discovered that yeast is not composed entirely of saccharomyces cervisiæ, but that other ferment-organisms, such as the ferment bacterium of acetic acid, or the bacteria of viscous fermentation are also fouud in it. These were separated, and a pure culture of the yeast obtained by the following method which he de-vised:

Into a test tube filled with sterilized grape juice, brewer's wort or other sacchariferous fluid, he introduced the small amount of yeast that would adhere to the point of a needle, and then sealed the tube against the admission of bacteria. In this tube fermentation would take place, and the yeast would grow and reproduce itself. A second tube was in-oculated from the first, a third tube from the second, a fourth from the third, and this transmission of yeast from one test tube to another, was continued until a "pure culture" of the yeast cells was obtained. It was correctly assumed that the yeast cells would grow and reproduce themselves more rapidly in the grape juice or brewer's wort, than would the other organisms, for the reason that these fluids furnish all the necessary conditions which this variety of ferment-cells require; whilst other ferment or-ganisms, those of acetic and viscous fermentations for ex-ample, would grow and multiply, but under the great dis-advantage of being less favorably placed than the yeast ferment; in consequence of this fact it was shown that on transferring a small amount of yeast from one culture-tube to another, and from this to a third, from this to a fourth, and so on, the number of organisms, other than the yeast fungus that would be also transferred, would become less and less, until finally there would be none, and a pure cul-ture of the yeast would be had. Pure cultures of other ferment-organisms were also obtained in the same manner. With these he was enabled to always induce a specific fer-mentation by adding the specific organism to the required fermentable sterilized solution and in this way proved that:

1st. Fermentation and putrefaction are caused by micro-organisms.

2nd. That a specific kind of fermentation is caused by a specific kind of ferment or micro-organisms: for example, the yeast fungus is the organism of vinous fermentation; the baccillus *acidi lactici* is the organism of lactic acid fermentation, the micrococcus aceti is the organism of acetic acid fermentation; and in the same way, each kind of fermentation has its specific ferment or micro-organism.

Now, this important work has not been fallow since Pasteur's announcement of his first work. He and many others have largely added to our knowledge of ferment bacteria, whilst bacteriology, as a science, has grown from infancy to the stature of vigorous manhood, and its methods and results are commanding the admiration of the scientific world.

This review of the history and phenomena of fermentation, enables us to point out some of the striking analogies which are found to exist between this process and that of infection. For example:

1.   Both fermentation and infection are caused by one-celled micro-organisms.

2.   These are particulate and portable substances.

3.   They are multiplied, and increase alike during the act of fermentation, and that of infection.

4.   In their increase, in both examples, they produce only their own kind.

5.   Both ferments and contagia are destroyed by over-heating, by antiseptics, and other similar agencies.

6.· In the same manner that ferment-organisms give rise to certain products during the act of fermentation, which products, when they have accumulated in a certain proportional amount, will arrest the fermentation; pathogenic bacteria, which are the contagia of contagious diseases, give rise during their action to certain products—ptomaines—which will arrest the disease when they have accumulated in the animal body in a certain proportional amount.

7. The kind of fermentation which will take place in any given case will depend largely upon the nature of the ferment; in other words, every fermentation is caused by a specific ferment.

8. The nature of any contagious disease will likewise be determined by the nature of the contagium; or in other words, every specific contagious disease is caused by a specific bacterium.*

9. Ferment organisms and pathogenic bacteria can be "attenuated" by certain methods; that is, their ability to do their specific work may be diminished, or totally destroyed; and this change, "attenuation," in the bacterium may be so firmly fixed that it will be transmitted through heredity from generation to generation of the bacteria, without in the least changing its natural appearance, even under the microscope, or its power of growth and reproduction. This is a remarkable fact, and I question if it has its parallel in organic development, unless it is found in other simple cells and low organisms. The yeast cells may be attenuated,—modified in their molecular structure, to such an extent that they will grow and multiply in brewer's wort, without causing fermentation, or producing a trace of alcohol. This has been accomplished by Oscar Breffield, and as a matter of fact, is carried on as an industry in the

---

*This rule is not absolutely accurate, perhaps, in either example, for certain fermentations may be caused by more than one kind of micro-organism; and on the other hand, it is probable that an infectious disease may be caused by more than one variety of bacteria; the general rule, however, is as above stated.

preparation German "barm." Other ferment organisms can be attenuated, modified in their specific power, in a similar manner. Pathogenic bacteria, those which cause infectious diseases, can also be "attenuated" or weakened in their power of producing certain products—ptomaines. Now, as these ptomaines are poisons, and are proven to be the cause of the symptoms and pathological lesions of infectious diseases, it can be readily understood that their virulence would be lessened by their attenuation, and that the disease or epidemic which they cause would be mild or severe in proportion to the virulence of its cause.

10. Hence ferment organisms and pathogenic bacteria can be modified alike by environmental causes.

CHAPTER II.

STRUCTURE AND PHYSIOLOGICAL ACTIVITIES OF CELLS—-
THEIR RELATION TO FERMENTATION AND INFECTION—
PASTEUR'S "RESPIRATORY THEORY" AND THE THE-
ORY OF "ENZYMES."

Having determined that fermentation, putrefaction and
infection are analogous processes, and are caused alike by
the growth and multiplication of one-celled micro-organisms in the fluids or substances acted upon, it remains for
us to inquire how these one-celled plants do this work, and
why one species or variety of cells will cause one kind of
fermentation or infection, and cannot produce other kinds;
for example, why the yeast cell can produce vinous fermentation, and cannot cause butyric fermentation.

Before entering into a discussion of these questions, and
the theories which they have developed, it is thought that
a description of the organisms in question, their construction, habits and physiology, together with a brief survey of
known facts regarding cells in general, will enable us to
better understand the wondrous mechanism of these microscopic laboratories, physical and chemical, called cells, from
which all living substances, all animal and vegetable organisms are built; they are the morphological units, the ultimate anatomical parts of all animal and vegetable forma-
tions.

The cell, according to those who first propounded the cell
theory, consists of a "cell-membrane," of a substance or
substances contained within the cell-membrane, called cell

contents, and of a central body or " kernel " called the "nucleus," differing in nature from the rest of the cell-contents.

More recent investigations have proven destructive of this idea of cell structure. It has been established that the nucleus is not necessary, and that the cell-membrane is sometimes absent, nothing of the cell remaining but the plasma, a very complex substance called protoplasm; and yet this naked, undifferentiated protoplasm will manifest all the functions of life, all the physiological properties of the highly specialized cell, but in a feeble and imperfect manner.

Some forms of amœbæ, so far as the most powerful microscopic lens can determine, are simply little masses of uncovered, undifferentiated protoplasm. They are living organisms without organs; they eat without mouths, digest and assimilate their food without digestive or assimilative organs, breathe without lungs, move without muscles, possess sensation without nerves (which is manifested by their contractility) and reproduce themselves without organs of reproduction. These simple protoplasmic units change their form, or their location, by pushing out a delicate stream of their substance, and then gradually flow towards it, or by extemporizing cillia, which they rapidly vibrate, and thus drive themselves through the fluids surrounding them; they flow around and envelop their food, and then convert it into living protoplasm. The rejected portions and the waste products of their bodies are dissolved and washed out by the liquid medium in which they live; they respire from their entire surfaces, take in oxygen and eliminate carbon dioxide; they react to stimuli by contraction of their substance; this often occurs independently of outside causes, as though it were the result of voluntary will power; and finally, they reproduce themselves by fission.

As may be inferred from this array of functions or conditions of protoplasm, it is not the simple homogeneous substance that all absence of internal differentiation of its parts into visible arrangements called structure would indicate; on the contrary, the many, and it may be diverse qualities of protoplasm, its metabolic qualities, indicate that it must have great complexity of molecular combinations.

If the difference in the character of work performed by apparently structureless protoplasm is due to a difference in the mechanical arrangements of its invisible parts, just as a difference in the character of work performed by a machine will depend upon the mechanical arrangement of its parts, then it would necessarily follow that it is the molecular condition of the protoplasm which gives this substance its protean qualities, and that different molecular arrangements correspond to the different forms of kinetic energy which protoplasm develops, or the•different kinds of work which it is capable of doing.

Protoplasm is the chief constituent of all cells, from the amœba to the highly specialized cell of the central nervous system of man. It is the universal life substance from which all organisms, whether animal or vegetable, originate. Professor Huxley has aptly called it "the physical basis of life." It certainly contains within itself, foreshadowings or imperfectly developed states of all the attributes of living matter, from that manifested by the simplest to the most complex cell structure, or from the protozoon to man. From its constructive and destructive metabolism, animal and vegetable growth derive all the energy of their bodies, whether manifested as heat, growth, movement, or other forms of physiological activities.

It is to the specialization of protoplasm that cells owe what Milne Edwards calls their division of physiological labor, whereby certain cells of an organism are set apart to do special work; for example, the labor of glandular cells is to secrete, of muscular cells to contract, and of brain cells to receive and distribute nerve impulses.

Protoplasm, then, is a body of many forms, a chemical substance of many isomers, a physical substance of many molecular combinations that differ among themselves and are as numerous as the stars in the firmament. In its simplest form it stands at the border line that divides the living from non-living substances, and separates the vegetable from the animal kingdom of nature. As to its origin, we know nothing,—except only that it is never created *de novo*, but originates from pre-existing protoplasm. It is a substance that is capable of undergoing, and as a matter of fact does undergo, great specialization which corresponds to, and increases *pari passu* with visible structural changes within itself which become most apparent in those cells whose aggregation and differentiation make up the different tissues of complex multicellular organisms. In many of the one-celled plants, and this is especially true of bacteria, the protoplasm, of which their bodies are entirely composed, is without visible structure;* hence the differ-

---

*Endogenous bacteria, during their sporulation, display a structural or visible differentiation of their protoplasm, which is intimately connected with their reproduction, but is not connected with the power which bacteria have of reducing organic substances; the exogenous varieties, for illustration, are equally possessed of this power, but do not at any time display a differentiation of protoplasmic contents.

ential work which they perform is not the result of those grosser changes which are denominated structure, but rather of the invisible activities of their molecular combinations.

It is not within the scope and purpose of this article to enter into a discussion of the progressive development of cell-structure, with its associated divisions of physiological labor, from unicellular to complex multicellular organisms; to follow the cells in their structural changes, in their specialization of functions, and the aggregation of the same kind of cells to form tissues, such as muscle, nerve, glandular, bone or brain tissue of animals, or those constituting the organs of the plant, however interesting this subject might prove. Some of these matters, however, will require notice, in order that a proper understanding may be had of the facts upon which are based some popular theories of fermentation and its allied processes.

Bacteria are microscopic plantlets composed of a single cell, which contains protoplasm; they are surrounded by a covering of cellulose, constituting the cell-membrane, and do not contain, so far as can be discovered, either a nucleus or other evidence of structural differentiation. Their number is beyond computation, their varieties are myriad and their size infinitesimal, often requiring the highest power of the microscope to bring them within range of vision. In shape they conform to three standards, viz.: a "billiard ball, a lead pencil, and a corkscrew." The first are called "cocci," the second "bacilli" and the third "spirilli." Their classification is based upon their growth and group-forms, and their method of spore formation. These subjects of primary classification are very important to the bacteriologist, but they do not especially concern our present inquiry, and we therefore will

pass them by and devote some attention to the sub-classifi-
cations, one of which is based upon the inability of bacteria
and ferment organisms to obtain their food from unorganized substances as do other vegetable organisms; the other
is the relation of bacteria to, and their effect on the sub-
stratum or food media, on which they subsist.

The fundamental constituents of living things are carbon,
nitrogen, hydrogen and oxygen. The comparative amount
of phosphorus and sulphur with different saline materials
which enter into the formation of living matter is trifling
when compared with the first named elements which make
up almost entirely the bulk of living substances. "Plants,
in their natural and healthy state, decompose carbonic acid
incessantly, fixing its carbon and setting free its oxygen.
Similarly they decompose water, seizing upon its hydrogen
and releasing its oxygen; whilst, lastly, they abstract nitrogen either directly from the atmosphere or indirectly from
the nitrate of ammonia which, under particular conditions,
has been formed therein. Plants, therefore, are marvelous
apparatuses of reduction, working with the aid of the heat
and light of the sun."

"Animals, on the contrary, are true apparatuses of com-
bustion. In their bodies carbonaceous matters are burnt
incessantly during the performance of animal functions, and
are returned to the atmosphere in the shape of carbonic
acid; hydrogen burnt incessantly is returned as water; while
nitrogen is ceaselessly exhaled in the breath, and thrown
off in the different secretions." "Thus we find that the
vegetable world is the great originator and source of that
pabulum which is necessary for the existence of animals.
Plants are the active agents ever ministering to the wants
of animals. Animals, as a rule, are powerless for the creation of organic matter; they can assimilate and modify the

organic substances which have been built up for them in
the tissues of plants; but they cannot extract from the
earth, air and water the elementary constituents of organic
matter, and force them to enter into such and such combinations."*

The power of tearing apart molecules of CO_2 –carbonic
acid,—the atoms of which are held together by powerful
chemical bonds, is possessed by leaves—the breathing
organs—of green plants, and is the result of sun light acting
upon chlorophyl, a green granular substance contained in
green leaves.   Now, those varieties of bacteria which
are connected with our inquiry do not contain chlorophyl,
and, consequently, cannot obtain their supply of carbon
from the atmosphere; in fact, they depend on organic compounds for food.

A popular theory of fermentation and allied processes is
based on the molecular and chemical changes in the
food media which result from the growth and reproduction of bacteria within its substance. I refer to
the respiratory theory of Pasteur.   During his investigations in fermentation, Pasteur early discovered that bacteria behave differently in the presence of atmospheric air,
or of free oxygen.   While all bacteria require oxygen for
respiratory purposes, some classes can take this directly
from the air and attain their full development and functional power only when thus placed; this variety he called
the aërobic bacteria.   Another variety, the anaërobic bacteria, have qualities which are directly opposite these; they
cannot thrive in the presence of air or free oxygen
and obtain their full development only when placed within
food media which does not contain these.   In this latter

---

*Beginnings of Life, by H. Charlton Bastian.

classification Pasteur places ferment bacteria and claims
they produce fermentation by robbing molecules of the
pabulum of a portion of their oxygen, and thus causing
them to fall apart. "He was led to this by careful qualitative studies of the alcoholic fermentation, and especially by
his discovery that the equation of Lavoisieur (see above)
cannot be regarded as correct. For in addition to the alcohol and carbon dioxide known to Lavoisieur, Pasteur found
always present small, though varying quantities of glycerine and succinic acid. But even with the addition of these
to the alcohol and carbonic dioxide, it appeared from his
analyses that about one per cent. of the sugar fermented
had disappeared. Out of one hundred parts of cane sugar,
about ninety-five parts are obtained as alcohol and carbonic
dioxide, three or four parts as glycerine and succinic acid,
while one part has disappeared. It is upon the loss of this
one per cent. of the weight of the sugar that Pasteur's 'respiratory theory' of fermentation is based; for he reminds
us that the yeast which does the work ought to profit by
the operation. It gains in weight and grows, and Pasteur
considers that it does so at the expense of the one per cent.
missing from the sugar. He also maintains that in doing
this, it is really breathing, and that in its hunger for oxygen, yeast attacks the sugar-molecule, and robs it of a portion of its oxygen, thereby causing it to break up into alcohol, carbonic dioxide, glycerine," etc.*

The scope of the respiratory theory is not confined to an
explanation of the hidden causes of fermentation. It is
claimed that the phenomena of contagion and immunity and
also the causes of these processes, result from changes, mo-

*Wm. T. Sedgewick in Reference Hand Book of Medical Sciences.
—Art. Fermentation.

lecular and chemical, in the fermentable liquid and blood, which the ferment or pathogenic organisms produce by withdrawing, for their own use, certain elementary portions from these fluids; when this results in a disruption and rearrangement of its molecular constituents, we have the phenomena and cause of fermentation, when it results in exhausting the fluid of those of its organic elements which are necessary to the life and growth of the micro-organism in question, we have that condition of the blood that renders it immune from the influence of the bacterium. Immunity thus secured, will, of course, continue until the organic elements which the bacterium has destroyed are again formed in the body.

This plausible and ingenious theory was received with marked favor and soon became very popular, partly on account of its analogies, its simplicity and seeming truth, and partly because it was propounded by Pasteur, that grand man to whom the world is so much indebted for original scientific work, which has greatly added to our knowledge of important scientific facts, has saved millions of dollars to France, and greatly added to her commercial prosperity.

Reasons have already been given why fermentation and infection are analagous processes, and it will be shown hereafter that immunity is intimately connected with and is a result of infection; hence the "respiratory theory" and all other theories claiming to explain the intimate nature of fermentation must be competent to explain also the intimate nature of its allied and associated processes. A discussion of Pasteur's theory will then necessarily lead us to inquire into its bearing upon fermentation, which will receive present notice, and upon immunity, which will be deferred to that part of this paper devoted to this subject.

If, as claimed by Pasteur, yeast cells cannot attain their

full development, morphological and physiological, except they obtain the oxygen which they require for their growth and multiplication, from organic molecules, then yeast could not thrive in grape juice, nor in brewer's wort, unless it could obtain the oxygen, which it needs for its growth, from the sugar-molecules which these fluids contain. Pasteur, in fact, claims that the yeast fungus is anærobic in its habit, that is, it can not receive oxygen from water or air, and cannot thrive in the presence of air or pure oxygen, it must, therefore, get the oxygen which it needs, from the breaking down of organic molecules; for example, the sugar molecules,—and he claims, furthermore, that the products of fermentation, such as alcohol, carbon dioxide, succinic acid, etc., are produced by a rearrangement of the molecules which have been set free. "The power of inducing vinous fermentation is not, however, confined to microscopic organisms; it has long been known from the experiments of Dobereiner and others, that sweet fruit, when kept in an inert atmosphere, devoid of free oxygen, evolves carbonic acid, with formation of alcohol, and it has been proved by Pasteur that this fermentation, which may extend to a considerable portion of the sugar present, is not accompanied by the development of any microscopic species. Closely related to this fact is the well established experience that large quantities of sugar may be made to ferment by means of yeast, without the latter multiplying to any great extent. On the other hand, large growths of yeast may be obtained, (and as a matter of fact, are obtained every day, by the makers of German barm) without producing much alcohol.

"Oskar Brefeld, by means of a peculiar artifice, succeeded in growing saccharomyces in brewer's wort, without producing a trace of alcohol."* And finally alcoholic fermentation may be caused by ærobic bacteria. These organisms not only thrive in the presence of air, but absolutely require to be thus placed; they obtain their oxygen from the air, or from water, and not from organic molecules.

The above enumerated facts, based upon trustworthy experimental proof, together with the equally well attested fact that yeast cells thrive, increase and excite vinous fermentation when they are freely exposed to atmospheric air or free oxygen, present some of the reasons why the "respiratory theory" has not received the support of many of the most eminent chemists, bacteriologists and biologists of the day. Additional and even more serious evidence of its incompetence will be presented when its relation to the phenomena and causes of immunity is examined.

We have presented in the foregoing pages a brief history of fermentation; have shown the striking analogy which it bears to contagion; the causative relationship which both processes bear to the growth of living micro-organisms or living ferments, and have presented the main features of the "respiratory theory" of fermentation, which is based upon, and is an outgrowth of our knowledge of living ferments.

We will now devote some time to a description of that class of ferments called non-living or unorganized, sometimes called enzymes, and to the theory of fermentation and in-fection which is based upon the origin and physiology of these bodies. Many of these ferments, and all those which are connected with our inquiry, are the products of certain specialized vegetable or animal cells which enter into the make-up of plants and animals. In following up the devel-

---

*Encyclopedia Brittannica, Ninth American Edition, Article Fer-mentation.

opment of cell-structure from the simple to the complex cell, we find that the functions of life, which are all manifested in the undifferentiated protoplasm of the simple cell, are divided and greatly developed in specialized cells whose protoplasm is differentiated into parts constituting structural arrangements. The tissues and organs of an organism are aggregations of cells of the same kind; thus the roots, the leaves, and the flowers, are plant organs, while the brain, the heart, the lungs, etc., are organs of animal structure. In each case the organ is made up of an aggregation of specialized cells of the same kind, which are especially adapted to perform the functions of the organ; while the organism as a whole consists in an orderly arrangement of its tissues and organs in a manner that best enables them to exercise their functions, give and receive required assistance, and conserve the interest of the entire organism.

The functions of life of a complex multicellular-organism are thus divided out, and performed by the specialized cells composing its tissues and organs. Specialized cells of each kind thus represent advanced stages of development of qualities, or functions, which cells have inherited from their ancient ancestor,—protoplasm. Thus the cells of the lungs, brain, glandular-system, etc., constitute specialists; organisms which devote themselves to one kind of labor, and have learned to do this very skillfully, at the same time, have left to other cells, and forgotten how, to perform other labors which are necessary to the plant or animal. "Each cell possesses its individual life, and passes through its particular course of development; it, however, does not undertake all the works of life, but limits the circle of its activities so as to reach a greater perfection within a smaller limit. In this it works, not for itself alone, but for the

other cells also, while it commits to them those requirements
for the satisfaction of which its individual activity is not
sufficient.   Thus the different functions are so divided
amongst the different cells, that one makes this, another
that occupation its own special business; one lives for all,
and all live for one."

Plants of the higher order of organization obtain their
food from air, earth and water, and their energy from the
sun; the roots absorb water and salts from the earth, which
are conveyed by the organs of circulation to every portion
of the plant's structure; the leaves of the plant dissociate
carbon from carbon dioxide, which is contained as an im-
purity in the air, they appropriate carbon, and set free
the oxygen, thus serving a double purpose in the economy
of nature—build up the plant's structure and purify the
air.   If the atmosphere was not purified it would soon become dangerously charged with carbon-dioxide and would
no longer be respirable.

The food material which the plant conveys into the system is carried to, and appropriated by each and every cell,
and by them converted into protoplasm.   This process is
called "constructive metabolism;" it comprises an act of
chemical combination that science has never yet been able
to accomplish by artificial means—that of making an organ-
ganic or living substance directly from the inorganic elements of chemistry   This process, which lifts dead matter
to the higher plane of living matter, requires a large ex-
penditure of energy which the plant gets from the heat and
light of the sun, and transforms into what is termed
"vital force" of the plant, whereby the plant-cells are
enabled to do their physiological work.   The protoplasm,
which the plant-cells thus form from the binary compounds
of chemistry is not a final product, but undergoes other

changes and is converted into other substances of complexity
—changes of a retrogressive character, whereby the relatively complex protoplasm is converted into products of a relatively simple structure. This process is known as "destructive metabolism," and is intimately associated with the
formation within the plant cells of unorganized ferments
which, in contradistinction to that other class called living
ferments, are denominated non-living ferments. They
are the products of certain specialized cells of vegetable and
animal structure; pepsin, trypsin, etc., represent the ferments which are produced by animal cells, and are formed
respectively by the cells of the gastric glands and pancreas,
those produced by vegetable cells are classified according to the nature of the chemical changes they induce.
The following embrace the principal ferments of this class,
viz: ferments that convert starch into sugar—diastatic
ferments; ferments that convert cane sugar into glucose;
—inverting ferments, also called "invertin;" ferments that
decompose glucocides, and lastly, ferments that convert
proteids, which are indiffusible, and, it may be, insoluble
in water, into peptones, which are both soluable and dif-
fusible. If these bodies perform a similar labor for plant
structure that those of animal origin do for animal structure,
then their function is to elaborate, digest and render assimi-
lable the food products on which plants subsist. The interest which we have in these substances is that especially
which relates to their modus in doing this work. We
know they require contact as living ferments do, and
that, like these, they add nothing from their own substance
to the ferment products and, if we can be guided by these
and other existing analogies, the similarity of phenomena
of fermentation lead us to believe that these are caused, in both cases, by similar if not identical agencies. But,

in fact, science knows nothing of that chemical composition
of ferments of either class by which they induce fermentation, and, as yet, no theory has been offered that can give
the rationale of this action and, at the same time, verify the
modus of the two kinds of ferments. The theory of fermentation which we will submit further on, we believe, will do
this, but we must not anticipate the explanation of this theory; other subjects must first receive attention. Having, for
the present, disposed of Pasteur's "respiratory theory," we
have briefly described the manner in which unorganized
ferments are formed in the cells of animal and vegetable
growths, as an introduction to a discussion of that theory
of fermentation which ascribes this and the closely allied
processes of infection and immunity, to unorganized ferments; these are also called enzymes, and the theory which
is founded on their action is, likewise, called the enzyme
theory.

"Invertin," it will be remembered, is one of the class of
unorganized ferments. Among the numerous sources from
which it is produced the yeast cell occupies an important
place, and it is well established that this ferment is the
agent by which cane sugar is converted into glucose, and
this conversion, or, more properly expressed, "inversion,"
is necessary to its fermentation and final conversion into
alcohol, carbon dioxide, etc. But, on the other side, it is
equally well settled that "invertin" has nothing more than
this to do with fermentation of sugar; yet, the enzyme
theory claims that all ferments secrete enzymes, as the
yeast cells do "invertin," and that these unorganized ferments are the true agencies or causes of fermentation. As
this theory also claims to explain the phenomena of infection and immunity, we must first examine its claim as a

theory of fermentation, and later on its claim as a theory of infection and immunity.

As a theory of fermentation, we believe it to be defective, first, in being not sufficiently warranted by facts, and second, it is directly controverted by experimental evidence. The first proposition is sustained by the fact that unorganized ferments are not found in fluids which have been fermented by living ferments, notwithstanding they have been examined for, by competent observers, time and again. It is evident then that they are not formed in this way, or, if formed, that they are destroyed so rapidly thereafter that they cannot be discovered by chemical reagents.

The second proposition is sustained by the fact, obtained by experimental work, that living ferments require immediate contact with fermentable substances before they can disrupt these. This is proved by the experimental work of Mitscherlich, Hoffman and others, which has already been stated; these experiments, it will be remembered, consisted in dividing the fermentable contents of a vessel into two parts by interposing a filter of cotton or bibulous paper between them, it was found now when a living ferment was placed in one of these, fermentation was confined to the same part of the vessel. Now, if living ferments secrete unorganized—soluble—ferments, as claimed by those who teach the enzyme theory, then they would be secreted by living ferments in the case referred to, and being soluable, they would not be limited to that part of the vessel which contained the living ferment, but would pass throught the filter, or animal membrane, as the case might be, and fermentation would, occur in like degree in both parts of the vessel.

Additional evidence against the "enzyme theory" is furnished by the fact that saccharomyces cerevisiæ can be made to grow and multiply in grape-juice or beer-wort without producing vinous fermentation. And, again, cane sugar may be inverted; starch may be converted into dextrine and dextrose. Salicine may be made to break-up into glucose and saligenine, and many other glucosides made to behave in a similar manner by the action of sulphuric and other strong mineral acids. The exact mechanism of these reactions is still unknown, but like the ferments before referred to, they act by their presence, and do not figure in the equations representing the chemical reactions which they induce.

In this connection it will be instructive to examine into the nature of that fermentative change which milk is made to undergo by the action of the bacterium of kefir. "Kefir," or "kephir," is the name of a drink, a fluid effervescing kind of sour milk containing a certain amount of alcohol, which the inhabitants of the upper Caucassus prepare from the milk of cows, goats, or sheep, and therefore is not to be confounded with koumiss, obtained by the Nomads of the Steppe originally from mare's milk; with this we are not concerned. The drink is prepared by adding to the milk the bodies described above as a beautiful example of the zoogloeæ, which bear the name of "kefir-grains." The Caucassians make use in this process of leathern bottles to hold the milk; the more polished European employs less objectionable glass vessels. The recipe followed by the latter is mainly as follows: Living and thorougly moistened kefir-grains are added to fresh milk, in proportion of about one volume of the grains to about six or seven volumes of milk. The mixture is exposed to the air for twenty-four hours, at the ordinary temperature of the room, protected from dust by a loose covering only, and is frequently shaken. At the end of twenty-four hours the milk is passed off from the grains,—which may be employed again for a fresh preparation. The milk itself—which we will term ferment-milk—is then mixed with twice its quantity of fresh milk, put into bottles well corked and frequently shaken. The bottled sour milk, which is more or less highly effervescent, is fit to drink in one or more days. It has the somewhat acid taste indicated by its name, and contains an amount of carbonic acid varying according to the temperature and duration of the fermentation, but sometimes sufficient to burst the bottles, or drive out the corks; and, as has been already said, a certain amount of alcohol, which, in the cases examined in Germany, was less than one per cent., but occording to other accounts, it may be one to two per cent.

The changes in the milk which produce the drink here described, are brought about by the combined activity of at least three ferment organisms. The kefir grains consist chiefly of the gelatinous filamentous bacterium, which has been named by Kern "Dispora Caucasica." Intermixed with this organism, and enclosed in the tough zooglœa are numerous groups of a sprouting fungus, a saccharomyces, resembling a yeast plant of beer; thirdly, there is the ordinary bacterium of lactic acid. We know at least enough of the ferment-effects of these organisms, or of their near allies, to enable us to form a probable idea of the course of the changes which have been described. The acidification is caused by the conversion of a portion of the milk-sugar into lactic acid by the bacterium of that acid. The alcoholic fermentation,—that is, the formation of alcohol, and a large part at least of the carbonic acid, is indebted for its material to another portion of the milk-sugar; and for its existence, to the fermenting power of the sprouting fungus.

The kefir grain, like its constituent, the sprouting fungus, working by itself, gives rise to alcoholic fermentation in a nutrient solution of grape sugar, though of a less active kind than that caused by the sprouting fungus of beer-yeast. But alcoholic fermentation is produced in milk-sugar as such, neither by sprouting fungi, with which we are acquainted, nor, as experiment has shown, by those of which we are speaking. To make this fermentation possible, the sugar must first be inverted, split into fermentable kind of sugar.

According to Nägeli the formation of an enzyme which inverts milk-sugar, is a general phenomenon in bacteria; and Hueppe has shown that it is probable in the case of his bacillus of lactic acid in particular. The inversion required in this case to enable the sprouting fungus to set up alcoholic fermentation, is the work therefore of the bacillus of lactic acid, or of the bacterium of the zoogloea, or of both.

" In this way we may explain the formation of kefir; and I gave this account of it in the first edition of this work, while calling attention to the want of precise investigation. *But A. Levy, of Hagenau, has recently discovered that the effervescing alcoholic kefir may be obtained without any kefir grains; simply by shaking the milk with sufficient violence while it is turning sour.* A trial convinced me of the correctness of this statement. The kefir obtained by shaking was not perceptibly different in taste or other qualities from the kefir of the grains. *Our former explanation therefore must be abandoned, and there is no other ready at present to take its place; but the case is full of instruction for our warning.*"*

----

*Lectures on Bacteria, by A. DeBarry. Second improved edition.

The description of kefir fermentation, and the explanation of its phenomena which this article furnishes, and especially that portion of it which we have placed in italics, contains matter worthy of thoughtful consideration; for if the views of its author are correct, that kefir grains are inverting ferments which prepare the sugar for its final fermentation by saccharomyces or the lactic acid ferment, and that the function which these grains perform may be accomplished equally well by violent shaking of the fluid while it is turning sour, then it is interesting to know the modus by which this result is obtained from such apparently dissimilar causes. The *prima facia* evidence is that in both cases the cause is mechanical, and that kefir grains, by their molecular activities, act on the sugar mechanically in a manner similar to that which results from violent shaking. If such is the case, then this illustration furnishes strong evidence in support of our physical theory, which will be explained later.

CHAPTER III.

MATTER AND FORCE—THE TEACHINGS OF MODERN SCIENCE
REGARDING THE NATURE OF MATTER AND THE "CORREL-
ATION AND CONSERVATION OF PHYSICAL FORCE"—ATOMS
—MOLECULES AND KINETIC FORCE DESCRIBED—A STATE-
MENT OF THE AUTHOR'S THEORY OF THE PHENOMENA
AND CAUSES OF FERMENTATION, THE ACTION OF MEDI-
CINES AND THE PHENOMENA OF CATALYSIS.

The foregoing review of the most important known facts
of fermentation, places within easy reach a means by which
the various theories of fermentation may be intelligently investigated and their values correctly determined. "Facts
are facts," to which theory must conform, and when it does
not, theory is at fault.

The final hypothesis of fermentation and infection to
which we shall invite your attention rests in molecular
physics, chemistry and biology. It claims that the laws of
matter and motion furnish energy to ferments, living and
non-living alike, that, in fact, vital force, so far at least as
it relates to physical action, is nothing more than transformed physical force, and is subject to the laws of "correlation"
and "conservation" that govern force in its physical manifestations.

Now, the subjects of matter, motion, energy, and "vital
force" will occupy a considerable amount of our attention
in this and the following article. We do not wish to anticipate what will be said on these subjects and bring them
forward out of their proper place, but we desire to briefly

state our case, i. e., to outline our theory before entering into a full discussion of it, and to do this it becomes necessary to state certain conclusions which, we believe, are logical deductions from the principles of matter, motion and energy. These are as follows, viz: energy, or dynamic force, the efficient cause of all physical phenomena, is derived from atomic, molecular or molar motion. The energy of a ferment, whether a living or non-living, is derived from the construction and motion of its molecules.

Spectroscopy teaches that atoms and molecules vibrate in periods of time that are distinctive of each kind of atom or molecule; therefore, the manifestations of energy by different ferments, i. e, their power of doing specific work, which differs with the kind of ferment, is a result of molecular grouping of their protoplasmic contents, and as protoplasm is a substance which has a complex molecular structure, of many isomeric forms, there will result a corresponding diversity in its manifestations of energy; for illustration, the protoplasm of an amœba is quite different from that of a bacterium, because of differences in the molecular grouping of the two.

Now, it is in that molecular structure, which gives a ferment its different forms of energy, we place the cause of its ferment power, and it is seen, therefore, that in some respects this theory resembles that of Stahl, and afterward, in a modified form, adopted by Liebig. This claims that fermentation and its allied processes are caused by molecular motions in the ferments; the nature of these molecular movements and the philosophy of their action are, however, explained differently in the two theories. "A ferment (according to Liebig's theory) is not a body *sui generis*, but rather any substance in a state of decomposition"--(Gerhard.) When such substance is brought in contact with

a fermentable substance whose elements are held together
by feeble affinity the more changeable substance, by virtue
of its inherent instability, will excite molecular movements
(motor decay) in even a large amount of the less stable
substance, and to this set of changes the name "fermentation" is applied.

Liebig's explanation is accepted by Gerhard, and is
thus described by him in his *Chimie Organique:* "Every
substance which decomposes or enters into combination,
is in a state of movement,—its molecules being agitated; but since friction, shock, mechanical agitation,
suffice to provoke the decomposition of many substances
(chlorous acid, chloride of nitrogen, fulminating silver)
there is all the more reason why a chemical decomposition,
in which the molecular agitation is more complete, should
produce similar effects upon certain substances. In addition,
bodies are known which, when alone, are not decomposed
by certain agents: thus, platinum alone does not dissolve
in nitric acid, but when allied with silver, it is easily dissolved; pure copper is not dissolved by sulphuric acid, but
it does dissolve in this when allied with zinc, etc. According to M. Liebig it is the same with ferments and fermentable substances. Sugar, which does not change when it is
quite alone, changes—that is to say, ferments—when it is
in contact with a nitrogenous substance undergoing change;
that is, with a ferment."

More recent investigations, and those upon fermentation
by Pasteur especially, have proven that "ferments" are
not, as claimed by Liebig, substances undergoing decomposition; on the contrary, the best known "ferments" are
living one-celled plantlets,—yeast cells and bacteria, and
we claim the molecular motions of ferments are not those
of decay (motor decay) but are motions of organization that

occur in periods of time that are definite for each and every molecular combination. Hence the theory which we respectfully submit differs from that of Liebig, not only in conception of the nature of "ferments," but also in the philosophy of their action.

As this theory is based on the laws of matter and force, its clear and intelligible exposition requires that some, at least, of these laws shall first be stated. They are as follows, viz: The indestructibility of matter; this is conceded by all persons who have given this subject a careful study. The seeming destruction of material substances is known to consist only in their transformation, and not in their destruction; the change is from one form or condition of matter into another form or condition; for example, ice may, by means of heat, be converted into water, then into an invisible vapor, and, finally, by passing steam through an intensely heated gun barrel the vapor is resolved into its ultimate elements, viz., oxygen and hydrogen.

Now, if these gases are collected into a receiver as they escape from the gun barrel, they may be recombined and again formed into water.

Scientific thought is tending more and more towards a recognition of the complemental doctrine of the essential oneness and indestructibility of force, and as force is regarded as a "mode of motion," and motion cannot be realized in thought except it be in connection with something that moves; it is believed that force, the efficient cause of all physical phenomena, is derived from the motion of atoms, of molecules or of mass; that attraction, light, heat, electricity, etc., are manifestations, or modes of one and the same force; that they are co-ordinate and are subject to the laws of transmutation of energy.

As we have stated, matter may undergo changes of form,

it may be now solid, now liquid, and now an invisible gas. Likewise force which is inseparably connected with matter —owing to such different modes of collocation of the atoms of matter—may manifest itself to us in different ways; first in the forces of crystalization (ice), then in those of liquids, then in the forces of vapor (steam), and finally, in the molecular forces of gaseous bodies. It will be observed that in these transformations of matter and force there is neither a destruction nor creation of either, simply a change of form of one, and a change of manifestation of the other. As Mr. Herbert Spencer says, "there cannot be an isolated force beginning and ending in nothing; but that any force manifested, implies an equal antecedent force, from which it it is derived, and against which it is a reaction. Further, that the force so originating cannot disappear without result, but must expend itself in some other manifestation of force, which, in being produced, becomes its reaction, and so on continually."

We have briefly presented some of the facts upon which is based our belief in the indestructibility and transmutability of matter and force; those grand generalizations in physics known as the "Correlation of Force," and the "Conservation of Energy" are founded on this theory.

Our discussion of these subjects has, so far, only related to gross matter and molecular force, but we will now ask your attention to the ultimate particles of matter—its atomic and molecular divisions—and to that force, or kinetic energy believed to be inseparably connected with, if not a primordial part of, these infinitesimal bodies. A molecule is the smallest division of a substance in which its qualities inhere; an atom is the ultimate part of an elementary substance. All substances, whether animal, vegetable or mineral, and all forms of matter, whether solid, liquid or gaseous, are built up of molecules composed of atoms. These latter are the foundation rocks of the Universe; they are indestructible and uncreatable.

Our block of ice may be regarded as a fair type of matter. It is sufficiently substantial and loaded down with properties, but a mere exposure to different temperature conditions causes its sensational properties to disappear. We soon come to a pair of invisible and intangible substances, oxygen and hydrogen, which can be investigated by indirect means only, but of which sufficient knowledge has been gained to establish their discontinuous or molecular character. "This molecule we must take as the representative of matter, for all masses of it, whether gaseous, liquid or solid, are but aggregations of similar corpuscles. We can only pursue it with the eye of the imagination, for its dimensions are so inconceivably minute as to far transcend the mechanism of vision. But could the molecule even be magnified to visible and tangible dimensions, with a new light to view it by, it could not by any means be rendered visible, either in whole or in its parts, on account of its incessant and marvelous activity, both interior and translatory. That the gas-molecule did not get its interior motion from the heat of dissociation is certain, for, in being allowed to recombine it yields up its translatory activity, and with it, as many degrees of temperature as disappeared in accomplishing the dissociation. No means of destroying the interior motion are known. By some scientists it is regarded as primordial and ultimate. We must not endow it with gratuitous attributes, but it is surely an entity of some kind, having in the first place persistent and regulated motion. Secondly, it has inertia, or mass, the property of conserving "*vis viva.*" Thirdly, it has some bond with its fellow, by which the motions of both are modified by a constant stress, accord-

ing to a definite law of distance; and this, following New-
ton, we call attraction. Fourthly, it has the complex property of interchange of momenta, accompanied by that of
conserving and compounding motion by angular rebound
upon an indefinitely near approach, which we name "resil-
iancy or repulsion." It is conceived as having dimension,
figure, polarity, elasticity and harmonic vibration with
"periods as undeviating as the moon and Mars.'"*

On this subject Professor Maxwell says: "Molecules
exist of various kinds, having their various periods
of vibration, either identical, or so nearly identical
that our spectroscopes cannot distinguish them. The same
kind of molecules, say, that of hydrogen, has the same set
of periods of vibration, whether we procure the hydrogen
from water, from coal, or from meteoric iron." Matter, then,
is made up of atoms These are hardly ever, perhaps never
found in isolation; two or more are bound together and do not
part company as long as the physical state of the substance
remains the same. Such elementary couples are called molecules. Faraday, in discussing the indestructibility of these,
says: "An atom of oxygen is ever an atom of oxygen; nothing in the least can wear it; if it enter into combination and
disappear as oxygen, if it pass through a thousand combinations, animal, vegetable and mineral; if it lie hid for a
thousand years, and then be evolved, it is oxygen with its
first qualities, neither more nor less. It has all its original
force, and only that; the amount of force which it disen-
gaged when hiding itself, has again to be employed in a
reverse direction when it is set at liberty." If, however,
the world was made of atoms and molecules alone, we

---

*An address by Henry Hobart Bates, A. M., and read before the
Philosophical Society of Washington, Jan. 27th, 1883.

should never know of their existence. · To explain the
phenomena of the universe, we must recognize the presence
of a continuous, universal medium penetrating all space and
all bodies.   This medium, which we call "ether," serves
to keep up the connection between atoms and molecules;
the motions of atoms are communicated to this medium and
by it are propagated through space as wave motions.
The wave motions produced in the ether by the harmonic
vibrations of molecules, recur in the same periods of time
as do the vibrations of the molecules which originated
them.   Thus, the molecular vibrations of an incandescent
substance will produce definite wave motions in the ether
which, when propagated through space, can, by means of
the spectroscope, be recognized and classified by their
periods, and the molecular constitution of the incandescent
substance can be determined by the same means. Spectrum
analysis, in truth, is based upon such facts, and by its aid
the molecular constitution of suns and other incandescent
bodies can be accurately determined, says Mr. A. E. Outer-
bridge, Jr. The delicacy of apparatus devised by physicists
and the refinement of demonstration rendered possible
thereby are among the greatest marvels of this wonderful
age.   The physicist is pushing his researches into paths
which but a few years since were thought to be forever
hidden somewhere in the vast realms of the "unknowable";
and the boundary line between so-called physical and met-
aphysical science is continually narrowing. Just as the
skilled mountaineer or the æronaut ascends gradually into
the rarified upper atmosphere, in order that the system may
accommodate itself to its new environment, so the philoso-
pher has advanced, step by step, until he seems almost to
have grasped the ultimate particles which constitute the
physical basis of the universe, and to render visible to mor-

tal unaided eyes particles of matter which are not only invisible by the aid of the most powerful microscope, but are too infinitesimal even for the mind's eye to conceive.

When that marvelous little instrument called the spectroscope was devised, it seemed that man had invaded fairyland and stolen "a trap to catch a sunbeam," for such it is, in very truth; not only does it catch the dancing sheaf of light, but it spreads it out into a band of exquisite colors and exhibits to our fascinated gaze Nature's palette of purest tints, out of which is woven the whole fabric of the gorgeous sunset, the varigated flowers, the bright plumage of birds, the iridescence of the mother-of-pearl, the sparkle and hue of gems, and, indeed every variety of color in nature or in art.

But this little instrument is still more wonderful, for it combines with its qualities of a trap the advantages of a balance which many suppose is fine enough for the most fastidious fairy to weigh the nectar distilled in the dew-drop, or other delicacy of the season.

Our ideas of weight and size are purely relative, and that which seems a small or light object, from one point of view, may become large and heavy by a different comparison.

To most of us, perhaps, a "grain-weight" suggests a little thing; we know that the apothecary, and a few other small dealers split up the grain into halves, quarters, and perhaps even hundredths, but then we regard them as homeopathic visionaries and laugh at their absurd little pellets; yet, strange to say, there is a vanishing point in our minds, which, if an object is small enough to pass, it becomes larger and more important by reason of our astonishment and wonder at its minuteness; the most ordinary specimen under the microscope is an evidence of this; but when we realize that the ability of the spectroscope to reveal small

particles of matter begins where the microscope searches
with its highest powers in vain, that the grain of matter may
be divided not only into hundredths, or thousandths, or tens
of thousandths, but into millionths and tens of millionths,
and that a single one of these particles may be detected by
this little searcher and held up for our inspection, our wonder and amazement enhance our respect for its occult powers. The astronomer tells us that a comet often throws out
a tail longer than the distance between the earth and the
sun, and broad in proportion; yet the matter forming the
tail is so attenuated that, if properly compressed, a gentle-
man's portmanteau, possibly his snuff box, might take it
in.* Yet we have merely to point this little tell-tale at the
comet, and, presto! we know what the matter is. Think of
it! Not merely can we grasp infinitesimal particles at our
hand, but we may sweep the firmament, gather up the star
dust and tell its composition.†

The most familiar waves are those produced in water.
Such waves have certain dimensions, e. g., a water wave
has a front, a back, a crest, a trough, and amplitude. The
crest is the top, and the trough is the bottom of the wave,
amplitude is the distance from the crest of one wave to the
crest of the next (nearest) wave.

Now it is a matter of common observation that water
waves may be influenced in certain ways by other water
waves. An increased size of the waves will result from one
influence and a diminished size from another (opposing)
influence. For example, when water waves meet from different directions there will result from the. union of such
waves an increase of their amplitude provided the approach-

---

*See Fragments of Science, Tyndal, p. 444.
†Popular Science Monthly, February, 1882.

ing waves coincide in their upward and downward movements, when, however, such waves do not coincide in
their periods,—if the wave troughs of one set correspond to
the wave crests of the other set, the amplitude of the resulting waves will be diminished, or, when the force of opposing waves is equal, the waves, are destroyed.

The law of interference may also be observed in waves of
sound, which are transmitted through the air, and in waves
of light, which are transmitted through the ether. Take,
for example two tuning forks which have equal periods;
when these are sounded each one will give out the same
continuous musical note. If, however, we change the vibrations of one of the instruments they will no longer
vibrate together, the vibrations will no longer coincide in
time, hence the sound that will be produced will be no longer a continuous musical note gradually fading away, but
a rising and falling sound. When the two instruments
vibrate together the sound will be distinct, but as
one vibrates more rapidly than the other they will gradually
part company and the sounds they give out will coincide
less and less in periodic time as one gains upon the other
and become fainter and fainter and disappear when the
wave-crest of one coincides with the wave-trough of the
other. After passing this line the sound becomes gradually
more and more distinct until it again reaches its acme when
crest and trough coincide with crest and trough of the other
side. Thus the sound will rise and fall depending upon the
amount of wave interference until the vibrations are finally
exhausted.

Light is also a mode of motion and, like sound has
waves which differ in their periods of vibration, but, unlike
sound, light waves are transmitted through ether. Illustration could be given that when waves of light are

met by other waves of light whose vibrations are one-half wave length out of time with those of the first, darkness will result; the waves of one set will quench or antagonize those of the other in the same manner that waves of water, or waves of sound, will interfere with other waves, of water or sound, whose periods of motion are out of time. Thus interfering waves of sound may produce silence, and interfering waves of light cause darkness.

The beautiful and important law of "interference," to the operation of which the above described phenomena are due, was discovered in the early part of the present century by Dr. Thos. Young, the author of the undulatory theory of light, and one of the most remarkable men in science and . literature of his century. The examples of its operation, which have been cited above, show that in its range of action it includes water waves, air waves and ether waves. Sir John Herschel, in speaking of the law, says: "This principle, regarded as a physical law, has hardly its equal for beauty, simplicity and extent of application in the whole circle of science."

In the light of modern science we can no longer regard matter as a "dead," inert substance. On the contrary, we must view it as having great molecular complexity, as possessing energy, and this energy or force, which is the efficient cause of all physical phenomena, is the result of and represents the total motions of its constituent molecules; light, heat, electricity, attraction and repulsion are simply "modes of motion," which, being subject to the laws of transmutation, are convertible one into another. When viewed in this light, what a marvelous piece of mechanism is the little microscopic cell about which we have had so much to say? We have already learned that it is composed principally of protoplasm; a substance of won-

drous complexity and protean qualities which, the teachings
of modern science lead us to believe, are the result of its
complex molecular structure. If it were possible to mag-
nify one of these little cells sufficiently to see the atoms and
molecules of its protoplasmic contents what a wondrous
sight would be offered us? How astonishing it would be to
witness the arrangements of its millions of molecules, and
millions upon millions of atoms all in constant motion, and,
under the rule of nature's laws, working harmoniously together for the common good. But if, through any device
of science or art, it were possible to witness the molecular
structure and movements of these bodies we would yet be
upon the threshold of the subject and would still have
much to learn regarding the organic life and history of the
cell. If all the facts of cell-life were known then life itself
would cease to be a mystery, it would become an open
book, because a knowledge of cell-life carries with it a
knowledge of animal and vegetable life inasmuch as animals and vegetables are simply aggregations, more or less
complex, of cells arranged and specialized in accordance
with natural laws to perform the various functions of the
organism. It may, then, be safely asserted that within the
compass of a little microscopic cell is concealed the mystery
of life.

Starting from modern concepts of matter, motion and energy, we can mentally follow the atoms in their chemical
combinations to form molecules, these to form other molecules, and these to form more complex molecules, and these
again to form still more complex molecules through an
almost endless series. And it is a reasonable supposition,
which is sustained by observation, that complex molecules
are more firmly united in some directions than others,
and that they are more readily separated along their

lines of weak union than along those more firmly fixed. When we add to this conception of atomic and molecular union, that of atomic vibrations in unvarying periods of time which are distinctive of each kind of atom, and that of etherial wave-motions vibrating in equal periods with the atoms that produce them, the law of "interference" enables us to understand how atomic wave-motions may be supplemented or antagonized by other atomic wave-motions, and how molecular wave-motions may, likewise, be similarly influenced by other molecular waves; that, in fact, the molecular waves which give a substance its energy will vary with molecular grouping. Now it is in these principles of molecular dynamics, and in chemistry and biology, that, we believe, is to be found the explanation of cell metabolism—constructive and destructive—of fermentation, of infection and of immunity. And in this connection we desire to express our belief that a correct application of these laws will solve many other dark problems of medicine, and that it will be along these lines of inquiry investigation will obtain correct knowledge of how medicines produce physiological action in the human body. It is well known, for example, that light, heat, electricity and sound,—which are conceded to be the modes of motion—have therapeutic properties, and are, in other ways, able to produce changes in the human organism, hence we feel warranted in the belief that medicines owe their therapeutic power to molecular vibrations, and, the therapeutic result which may be obtained from the use of electricity, or from the administration of a drug, is, in both cases, *caused by wave vibrations.*

We have now reached the stage of our subject when we can intelligently answer the inquiry how little microscopic cells called ferment organisms cause fermentation, and why

one species or variety of cells cause one kind of fermentation and cannot cause other kinds? For example, why yeast-
cells produce vinous fermentation, and cannot cause butyric,
acetic or lactic acid fermentation? In elucidating the occult
causes of fermentation we will refer, for illustration to the
action of yeast-cells on grape juice in the production of alcohol, carbon-dioxide, etc., as this is the most familiar and
best understood fermentative process. At the same time
the principles involved in the explanation of this process will apply with equal force to an explanation of all
other varieties of fermentation whether caused by living, organized ferments, or by non-living, unorganized ferments.
They will also serve to explain how pathogenic micro-
organisms give rise to infection, the laws of infectious diseases and, as well, the phenomena and causes of both natural and acquired immunity.

When viewed in the new light of science, our little yeast
cell is seen no longer to be the simple body that its gross
appearance and seeming absence of structure would indicate, but is regarded as an instrument of great complexity in its mechanism, with energy or power to do certain
work.

The terms "instrument" and "mechanism" which we
have applied to bacteria cells, must not be understood as
conveying the idea that these have visible parts which are
arranged in accordance with mechanical laws similar to
those observed in mechanical contrivances. On the contrary, these cells are without visible structure, they are
apparently microscopic specks of protoplasm enclosed in
enveloping membranes, and yet there is a similarity both
in motive power and construction between the two. Take
for illustration a large number of differently constructed
machines driven by the same engine, one machine

will perform work of one kind because of its mechanical construction, another will perform different work, and others will all do that kind of work for which they were specially constructed. In this case, it will be observed, the power that drives these machines is the same for all, while the kind of work performed will vary with the construction of the "machine." Now, the engine gets its power from the coal it consumes, the coal its power from former vegetation, while this obtained its power from the sun—the supposed source of all physical force,—but if we carry our investigations a little further, we will find that the sun gets its force from the atomic and molecular vibrations, interior and translatory, of its component parts. In the other case, the bacteria cells obtain their force, or energy, from the same universal source, atomic and molecular vibrations, which they derived, in part, from the light and heat of the sun. Again, the character of work which a bacterium can do depends on the arrangement of its invisible parts as that of a "machine" does on the arrangement of its visible parts.

We will now examine the subject of fermentation in the light furnished by these principles of molecular dynamics, and will also determine the ability of our new (physical) theory to philosophically explain the intimate nature and phenomena of this process. In addition to those characteristics which are common to all protophytes, such as protoplasmic contents, cell covering, organic and functional life, we must regard the yeast-cell as having a molecular structure that is definite and characteristic of such cells, that this structure produces etherial wave motions of equal periods with its vibrating molecules, and that from these the cell obtains its specific energy; i. e., its power to do specific work. The analogy between a living cell and a mechanical contrivance is again brought out by this theory of cell action, in the same manner that a machine derives its special energy,—its ability to perform specific work,— from the construction and mechanical arrangement of its parts, we claim the cell derives its ability to do specific work from the arrangement,—structure, — of its invisible parts—atoms and molecules. For example, yeast cells and others which comprise ferment bacteria, owe their definite form of energy—their power of inducing that form of fermentation of which they are the respective cause to the distinctive grouping of their atoms and molecules.

The distinctive energy or waves of a cell, say a yeast cell, can influence those substances only whose waves bear a certain relationship to those of the yeast cell; they must be equal in their periods, direction and, perhaps, in other characteristics, before those on one side can influence those on the other. The nature of this influence will again depend on whether the two sets of waves coincide in trough and crest. If they do, the waves will supplement each other and their amplitude will be enlarged; if they do not, they will antagonize each other and their amplitude will be diminished, or, it may be, the waves will be destroyed by mutual antagonisms; it will be remembered that all this actually occurs in waves of sound, of light and of water, and if analogy has any merit, it can also occur in waves of molecular energy.

Now, if grape sugar has waves whose periods equal those of yeast, and the crest and trough of both coincide, it follows, that when the mass contact of grape sugar is destroyed by dissolving this, say in a fermentable solution, and to this solution is added yeast—the vinous ferment— the waves of the yeast, falling many million times in

a second upon those of sugar molecules will increase
the amplitude of these, which, in being driven back upon
their atomic constituents, will force the atoms further and
further apart, until they will be forced finally beyond their
chemical bonds and the sugar molecules will be disrupted,
that is they will be resolved into their atomic condition. It
is evident that the liberated atoms cannot long remain in
this free condition, chemical law will speedily force them
into other combinations. This is really what takes place
in vinous fermentation, and alcohol, carbon dioxide, etc.,
are the substances which are formed from a recombination of the atoms of the disrupted sugar; for chemical analyses prove that these products, collectively, contain the
same atoms and in the same proportions which the grape
sugar contained.

In this connection we wish to call attention to an important phenomenon of fermentation which, as a matter of
fact, is well established, but its philosophy and mode of
action remain to this time a profound mystery. I refer to the
unquestioned fact that the products of a fermentation inhibit the action of the ferment, and that an accumulation
of these will finally arrest the fermentation. Now, this
fact, even if it had not been known, could have been foretold
and its philosophy and cause determined by reasoning inductively from the premises on which our new theory
rests; in truth, this fact is a corrolary of our theory, as the
following explanation will prove. Assuming that fermentation is produced by two factors, viz., a ferment and a fermentable substance, whose molecular waves coincide in time
and periods, and that the products of fermentation are
derived from a recombination of the atomic constituents of
the fermentable substance which was disrupted by the molecular bombardment of the ferment, it is apparent that the

ferment products which can form against this influence must have molecular waves that can not be influenced by those of the ferment, for it is evident that the same influence— which continues to act---that disrupted the fermentable subtance will surely prevent the formation of another having equal vibrations. It follows, therefore, that the products must have waves which do not coincide in time with those of the ferments, i. e., which antagonize these, and will therefore inhibit their action. Such being the fact, an accumulation in sufficient quantity of these products will arrest the fermentation.

Thus, the yeast cells cause vinous and can not cause other fermentation; for example, that of butyric, acetic or lactic acid, for the reason that the molecular vibrations of these cells do not recur in periods of time with those cells which are the special ferments of these processes. It is only when the molecular vibrations of a ferment, whether this be a living, organized ferment, or a non-living, unorganized ferment, coincide with those of a fermentable substance, that the latter may be disrupted by the former, and fermentation ensue. The reason why ferments act by their presence, produce fermentation without figuring in the resulting chemical equation, becomes at once apparent when viewed in this way.

Light is also thrown upon the dynamics of catalysis; why certain substances can, by their mere presence, decompose compound substances; e. g., the decomposition of hydrogen peroxide by the presence of spongy platinum, remains to this day an unsolved problem. The process is called that of catalysis, and the cause is ascribed to an "occult, special, unique and hypothetical force."—(Thomas.) *How simple the explanation of catalysis when the "catalytic agent"*

*is known to be a substance whose molecular vibrations shake
in sunder the molecular combinations of the other substance!*

The "molecular theory," in its new aspect, serves also
to explain why "platinum alone does not dissolve in nitric
acid, but when allied with silver, it is easily dissolved; why
pure copper is not dissolved by sulphuric acid, but is dissolved in this if allied with zinc;" and, why "friction,
shock, or mechanical agitation suffice to provoke the decomposition of many substances (chlorous acid, chloride of
nitrogen and fulminating silver); aud, in the manufacture
of kefir, why brisk shaking of the milk while it is turning
sour, will produce those changes, which DeBarry at one
time taught occurred only when a special "enzyme" "inverted" the milk sugar. The fact that the fermentative process occurs only when the fermentable substance is in immediate contact with the living ferment, and that a piece of
filtering paper, or an interposed thickness of cotton is sufficient to confine this act to the portion of the vessels which
contains the ferment, has proven a serious stumbling block to
the enzyme theory, and, as well, to the respiratory theory;
but when the ferment cells are regarded as physical agents,
and fermentation as the result of kinetic energy, these facts
become strong evidence in favor of our physical theory.

CHAPTER IV.

ETHERIAL WAVE MOTIONS AND ILLUSTRATIONS OF THEIR
WONDROUS ENERGY—AN EPLANATION OF THE CAUSES
AND PHENOMENA OF INFECTION—THE PRODUCTION OF
TOX-ALBUMINS OR PTOMAINES—THE MOLECULAR NAT-
URE OF ALBUMINOIDS, AND THE SELF-LIMITED NATURE
OF ACUTE INFECTIOUS DISEASES.

It may appear, at first glance, that too much energy is
ascribed by our physical theory to etherial vibrations which
are produced by bodies so infinitely small as atoms and
molecules; that the chemical bonds which unite these elements into compound substances are too powerful to be
overcome by forces which are apparently so feeble.

The potency of these vibrations will be recognized, however, when we remember that these waves recur in set periods of time; are rapidly followed by others of equal periods, and that wave impacts are made many thousand times
in a second. Therefore the force of a single impact must
be multiplied by thousands to determine the energy produced by these waves in a second of time. It must also be
remembered the molecular vibrations of any two substances
must bear a certain relationship, regarding their periods of
vibration, in order that the waves from one set may influence the waves of another. In a similar manner, the arc
described by a moving pendulum will be increased or diminished by air waves; increased when the pendulum and
the air waves move in the same periods, and diminished
when these periods are different. In fact, if we credit the

recent teachings of science, then atomic, molecular, and molar motions, with their associated ether-vibrations, constitute nature's great store-house of energy; and, in whatever manner this energy is manifested, whether it be in the form of light, heat, or electricity, it is certainly wave motion and nothing more; the difference in its manifestations or modes being only a difference in wave lengths, or periods of the ether-vibrations.

The vast importance of ether wave motions in the causation of physical phenomena, their bearing on modern science, and their causative relation to fermentation, infection, and allied processes especially, leads us to believe the following quotations will not be considered out of place. Prof. Tait says: "It is very difficult to realize the fact, certain as it is, that light (in the sense of brightness) is a mere sensation, or subjective impression, and has no objective existence. Yet we know that, besides those radiations which give us the sensation of light, there are others, in endless series, both higher and lower in their refragability, to which our eyes are absolutely blind. And the only difference between these and the former is one of mere wavelength or period of vibration.

"Similarly, it is very hard to realize the fact that sound (in the sense of *noise*) is only a sensation, and that outside us there is merely a reries of alternate compressions and dilatations of the air, the great majority of which produce no sensible effect upon our ears."—[*Encyclopædia Britannica,* *9th ed., art. Mechanics.*

Prof. Wm. Crooks says : "It has been computed, that in a single cubic foot of the ether, which fills all space, there are locked up ten thousand foot-tons of energy which have hitherto escaped notice. To unlock this boundless store and subdue it to the service of man, is a task which awaits the electrician of the future. The latest researches give well-founded hopes that this vast store-house of power is not hopelessly inaccessible. Up to the present time, we have been acquainted with only a very narrow range of etherial vibrations, from extreme red on the one side to ultra violet on the other—say from three ten-millionths of a millimetre to eight ten-millionths of a millimetre. Within this comparatively limited range of etherial vibrations, and the equally narrow range of sound vibrations, we have been hitherto limited to receive and communicate all the knowledge which we share with other rational beings. Whether vibrations of the ether, slower than those which affect us as light, may be constantly at work around us, we have, until lately, never seriously inquired. But the researches of Lodge in England, and Hertz in Germany, give us an almost infinite range of etherial vibrations or electrical rays, from wave-lengths of thousands of miles down to a few feet. Here is unfolded to us a new and astonishing universe—one which it is hard to conceive should be powerless to transmit and impart intelligence."—[*Popular Science Monthly*, February, 1892.

Additional evidence of the marvelous energy of etherial wave-motions is furnished by an examination of sunlight. The accepted theory of light is that it consists of etherial wave-vibrations given out by an incandescent body. Now, the sun is an incandescent body, and the atoms and molecules of its composition are in a state of marvelous activity, which is caused, in part, by the intense heat of this incandescent body, and in part by the primordial motions of the atoms and molecules. These motions, which are both interior and translatory, convey wave vibrations of definite periods to the surrounding ether which furnish light and heat to the world. When a narrow pencil of white light from the sun or other incandescent source is passed through a glass prism, the light is refracted and forms a lengthened spectrum which is made up of the seven colors of the rainbow, viz: red, orange, yellow, green, blue, indigo and violet, and when this spectrum is examined by a magnifying lens of sufficient power, it is found to be crossed by many delicate lines which are arranged in uniform order, and always occupy the same parts of the spectrum.

It would prolong this article needlessly to present the indisputable evidence on which is founded the belief that each of these lines, or group of lines, is produced by molecular vibrations in the incandescent body from which sun light is obtained. Thus sodium has its lines which always occupy the same place in the spectrum; likewise, oxygen, hydrogen, and the other elements have their characteristic spectra by which their presence in the sun may be positively determined. The solar spectrum contains other rays or wave vibrations which, although invisible to man, are quite necessary to his existence. For example, the presence of heat waves, although invisible, can always be demonstrated at the red end of the spectrum, and at the opposite (violet) end, other invisible waves are always found, which, because of special qualities, are called actinic or chemical rays.

Sunlight, then, contains heat waves, color waves, actinic or chemical waves, and those waves which give rise to the cross lines of the spectrum; all of which, when traced to their source, the sun, are found to be a result of atomic and molecular vibrations of that incandescent body. At one end of the chain of causation is the sun, an incandescent body whose ultimate particles are vibrating in active and periodic time; at the other end is the earth. Connecting the two, the sun and earth, is a universal medium which receives the molecular vibrations of the sun and transmits them to the earth where they may be separated, investigated, and made to tell their history and origin by means of the spectroscope. Now these minute wave-motions travel millions upon millions of miles, and require ages to make their transit from the sun to the earth, yet they are transmitted by the universal ether with even greater accuracy than are the vibrations of the human voice with all the tones, modulations and inflections of the speaker's voice by · a telephone wire from a distant telephone.

When it is considered that these wave-motions constitute an important part of nature's forces, and that it is through these agencies that she is enabled to perform her manifold and wonderful work and, that through similar agencies we seek to explain the phenomena of fermentation and infection, it cannot be justly claimed that the causes assigned are inadequate to perform this work.

Coming back to the main subject of discussion, it will now be shown that the physical theory will explain why pathogenic bacteria cause infectious diseases, why these diseases are infectious, why this class of diseases are self-limited in their duration, and the causes and philosophy of immunity, as readily as it does the causes and phenomena of fermentation. While the conditions of fermentation are, in many respects, quite different from those of infection; e. g., one process may involve only two non-living substances and is terminated by the formation of ferment products; the other involves two living substances—the bacterium and the tissues of a complex multicellular organism,—and the process is not terminated with the formation of infectious products, yet the physical principles which, we believe, are the active agencies of these processes are the same in both cases. In the same manner that ferment bacteria disrupt and convert fermentable substances into ferment products, we claim that infectious bacteria disrupt and convert albuminoids into infectious products or tox-albumins. The causative relationship which bacteria bear to this class of diseases is too well established to require a defense, suffice the statement that in many of these diseases the bacterium, which is the undoubted cause, can be always found in the infected individual, it can be transferred to culture media and grown in pure cultures, and bacteria taken from these cultures and inoculated into the bodies of susceptible animals will invariably produce the original disease. It would seem that, in the face of this evidence, there can no longer be reasonable grounds for doubting that bacteria are the efficient causes of these diseases. The questions requiring answers are those which relate to the manner in which bacteria act, and the causes of the phenomena of infectious diseases, which we will now give.

It will be remembered that ferment bacteria must find in fermentable fluids, substances which they can disrupt and convert into products of fermentation before they can display their peculiar action. It is the same with pathogenic bacteria or those which produce disease; they must find in the body-juices substances which they can shake apart and convert into ptomaines, toxines or tox-albumins before they can display their pathogenic properties. It is generally conceded that the substances upon which pathogenic bacteria thus act, are the "albuminoids." These substances, like protoplasm, are of complex molecular structure, and are very unstable in their chemical combinations. In fact, the term albuminoids comprises many different substances which are, perhaps, of the same chemical composition,—that is, contain the same "elements" in the same proportion—but possess many differences in the manner these elements are grouped together; in other words, while the kind and number of molecules which albuminoids contain may be constant, their manner of grouping, which constitutes structure, is subject to great variations. The necessity for this difference in the molecular structure of albuminoids becomes more apparent, when we remember that they are derived from the food which has been eaten; that this food has been changed by the act of cooking, and further changed by the action of the digestive ferments; that they supply to all the cells of the body (bone-cells, nerve-cells, etc.,) the nourishment which they require for their growth and repair, and finally, that this substance is converted by these cells into their own structure. A substance of uniform molecular structure would hardly be able to meet the demand which the cells of the body make upon the albuminoids.

There are other conditions of molecular structure of albuminoids which have an important bearing upon our physical theory. For example, a molecule of albumen is not a simple body like that of water; it is believed to be a very complex body composed of many molecules that have lines of weak union, like the lines of cleavage in crystals, along which it is most easily broken. These important bodies which occur in all the tissues of the higher animals and form their principal bulk, and are so necessary to the nutrition and growth of animals may, under abnormal conditions, be converted into exceedingly poisonous substances. The nature of the poison, which is thus formed, will depend along what lines of weak-union separation takes place; and this, again, will be determined by the nature of the disrupting force.

"At least three distinct series of chemical bodies are thus formed, viz.: an acid series, an aromatic series, and a basic series. Out of the innumerable products arising from the action of bacteria cells upon albuminoid molecules, and which have been extracted and studied, will be mentioned indol, cresol and skatol in the aromatic series; creatine in the basic, and uric acid in the acid series, serving only as mere examples."—*Jour. Am. Medical Association.*

Dr. Brunton, in his address on medicine before the last session of the British Medical Association, in discussing this subject said : "Albumin might be compared to glass. Split it in two and it becomes harmful, join the broken halves and it becomes harmless again. Microbes might have this action on albumin in the organism." The albuminoids, then, which exist largely in the blood and body-juices of animals, are the substances which are disrupted and converted into poisonous albumins—tox-albumins and tox-ines—by the molecular vibrations of pathogenic bacteria. The nature of the substances formed in any given case will depend, first, on the specific bacterium concerned,—its molecular structure and resulting wave vibrations,—and, second, the albuminoid concerned,—its molecular structure and resulting wave vibrations.

Any bacterium is pathogenic to an organism when the molecular wave vibrations of this bacterium coincide in periods and time with those of an albuminoid of the organism and the former can disrupt and convert the latter into poisonous albumins. Therefore, when there are a multiplicity and diversity of molecular groupings in bacteria, and a similar multiplicity and diversity of groupings in albuminoids, there will result from the action and reaction of these bodies—one upon the other—a multiplicity and diversity of products, tox-albumins, etc. Hence, bacteria manifest a selective action; one variety will disrupt certain albuminoids, another variety other albuminoids, and so on through an almost endless series. The resulting products will vary in each case; that is, each variety of bacteria will produce its characteristic tox-albumins.*

Bacteria are not, however, the only bodies that act in this manner. It has been stated* elsewhere in this essay, that living ferments excite fermentations in the same manner that pathogenic bacteria produce infection; and it is known that digestive ferments also decompose albuminoids and convert them into poisonous albumins.

----

*Abstract of a paper by Edward O. Shakespeare, on "Investigations Concerning Poisonous Substances Produced by Bacteria." By L. Brieger and Carl Frankel, of Berlin. Published in the Berliner Klin Wochenschr, No. 11 and 12, 1890:

"Thsee authors relate the details of a very thorough research upon the chemical products to be found in the cultures of the Lœfler bacillus of diphtheria, of the typhoid bacillus, of the anthrax bacillus, of the tetanus bacillus, of the cholera bacillus, and of the staphylococcus aureus. For a full account of these researches the reader must consult the original paper.

"In bouillon cultures of the Lœfiler bacillus of diphtheria, they have found a substance which, by its chemical constitution and behavior, should be classed with the albuminoids, and which possesses marked poisonous properties. According to the results which they have obtained, they think that they are not unwarranted in expressing the opinion that this albuminous substance plays a very important role through the characteristic action upon the organism of the bacillus diphtheritical of Lœfler. They hold also that they have found other bacterial products, of a similar constitution and behavior, to which they—in contradistinction to toxin—give the name tox-albumin, are elaborated, and that these latter substances appear to be even more important than the former.

"Within the living body this tox-albumin is doubtless built up out of the albumin of the tissues and decomposed again by which the latter, through a transposition and alteration of its atomic groups, acquires poisonous properties. For the demonstration of this, however, further experiments must be undertaken.

"In their artificial cultures, the tox-albumin certainly originates

An editorial in the *Journal of the American Medical Association* says: " Cells other than those of bacteria are also capable of decomposing albuminoid-molecules, and thus produce poisonous substances; c. g., in the physiological changes from albumins to peptones, there is a change from innocent to toxic bodies. This may be illustrated by the hypodermic injection of the digestive leucomaine of fibrin by pepsin. It occurs practically when, after a too hearty

---

from the blood serum which is added. These authors, it need not be said, tested the nutrient fluid for a poisonous substance which might have been formed by the numerous processes through which it had gone before inoculating it with the bacteria, but with the negative result. In those cultures which contained no serum, but in which, nevertheless, a poisonous substance was elaborated, this could only have been produced by the peptone which had been undoubtedly converted back again into albumin.

"SUMMARY.

"These authors have, to use their own language, thus found that an entire series of different micro-organisms, among them the most infectious, which, in their artificial cultures and the bodies of animals killed by them (experiments with anthrax bacilli), elaborated substances that, according to their chemical qualities, are to be regarded as direct derivatives of albumen, and that possess such a decided poisonous character that we have given to them the name of tox-albumin. These authors believe that these substances, through the noxious action of the pathogenic bacteria, play a very important role, and occasion the ordinary symptoms, and, under some circumstances, also the death of the animal attacked. Since the tox-albumin is elaborated in the body of the animal, doubtless from the albumen of the tissues, and does not fundamentally differ from it, so the latter acquires an enhanced interest in pathology. The distance from the normal constitution of the body to substances of the dangerous kind appears shorter than we formerly imagined, and bodies themselves seem to be the immediate cause of the morbid conditions which are brought about through the vital processes of bacteria.—Report of Cholera in Europe and India.

meal, the liver is unable to care for the excess of digestive
leucomaines, and they escape into the circulation, producing somnolence, lassitude, or even stupor.''

On the same subject Dr. Joseph Leconté says: '' The
leucomaines, although formed by normal physiological process, are highly poisonous and inimical to health, unless
speedily eliminated by appropriate organs. If, now, there
should be a failure to eliminate these toxic elements, the
result would be similar to those produced by disease germs,
except that they would lack the quality of contagiousness,
because they are not due to the presence of microbes. The
liver is the organ principally concerned in the elimination
of leucomaines: if alkaloids, by the conjunction of S. biliary
acids; if carbo-hydrates, they escape the liver and are taken
care of by the blood and pancreas. If belonging to the
class of phenols, they are combined with a sulphuric radical, and when that gives out, are then combined with gly-
cosuric acid, and thus rendered innocuous.''

The statements contained in the above quotation give
prominence to, and strongly support the views which have
been urged with so much earnestness in this essay, viz.:
that the *modus operandi* of ferments—whether organized
or unorganized—and infectious bacteria, is the same.
But the similarity between digestive ferments and pathogenic bacteria ceases to exist at this point; there is a
radical difference in their nature, which carries with it an
important and practical distinction. I refer to the fact that
pathogenic bacteria are living organisms which multiply
themselves by the generative act, and are therefore contagious elements, capable of giving rise to, and spreading,
contagious diseases. On the other hand, digestive ferments are simply molecular combinations without organic

life, consequently have no power of self-multiplication, and cannot cause infectious or contagious diseases.

In speaking of pathogenic bacteria as disease-producing organisms, I do not wish to be understood as saying that these bacteria are the immediate cause of infectious diseases, for such is not the case; their role of action is to convert albuminoids of the body into poisonous albumins, and these products, not the bacteria, produce the symptoms and pathological lesions of the disease.

"The science of bacteriology is now entering upon a new and promising era in its development. Heretofore this science has been largely founded upon morphological studies. Bacteriologists have given their time and attention to the discovery of bacterial forms in the diseased organism, and to the observation of characteristics in structure and growth of different species of bacterial life. The question, Do certain germs have a causal relation to certain diseases? having been settled in the affirmative, the next question which naturally arises is, In what way is this causal effect accomplished? How do germs prove harmful? To this question a number of answers have been proposed. It is now generally admitted that the deleterious effects wrought by germs are due to chemical poisons elaborated by them during their growth. Granting this, it will be at once seen that the morphological study of germs, important as it is, becomes wholly inadequate in ascertaining their true relationship to the diseases with which they are associated. We must now study the chemistry and the physiology of the germs, and until this is done we must remain ignorant of the true cause of disease; and, so long as we remain ignorant of the cause, it cannot be expected that we shall discover scientific and successful methods of treatment. Suppose our knowledge of the yeast plant was limited to its form and method of growth, of how little practical importance this knowledge would be. That the yeast plant requires a saccharine soil before it can grow, and that given such a soil, it produces carbonic acid gas and alcohol, are the most important practical facts which we have ascertained in its study. Likewise, the conditions under which pathogenic germs multiply, and the products which they elaborate in their multiplication, must be ascertained before their true relationship to disease can be understood."*

The correctness of this view is proved by observation and experimental work. For ¡illustration, two, at least, highly infectious bacteria, those of diphtheria and tetanus, are strictly ærobic and cannot live within the animal body; they are confined to its surface, where they elaborate their poisonous albumins, which, being soluble, are carried by the blood and lymph circulation throughout the body. Further proof is furnished by the fact, that infection of the body can be induced by inoculations of the purified products of bacterial action which have been obtained from pure cultures of the bacteria; or, when these can not be obtained, the same result will follow an introduction into the body of the fluids of such cultures after thorough separation of the bacteria. This conversion of the albuminoids into other forms of albumin—tox-albumin—by the dynamic energy of bacteria, is termed destructive metabolism; wherein a relatively complex substance is converted into one that is relatively simple, and as this process always causes an evolution of heat, it is doubtless to such metabolic changes that the increased temperature which

---

*Chemical Factors in the Causation of Disease, by Prof. Victor C. Vaughn. Journal American Medical Association— Bacteriological World.

attends infectious diseases is to be attributed. The tox-
albumins of pathogenic bacteria have been extracted from
pure cultures of many of these organisms, and it is known
that inoculation of these purified products into a susceptible animal will result in producing that disease which may
also be produced by inoculations of the bacteria from
which they were obtained. Tetanus, diphtheria and pneumonia of animals are cited as mere examples.

Attention has already been called, in that part of this
essay which treats of the intimate nature and phenomena
of fermentation, to the fact that products arising from the
action of ferment organisms upon fermentable substances,
will bring the fermentation to an end, and, that it can not
be re-established until the products, or at least a large part
of them, are removed, when fermentation will again take
place and continue until again arrested by an accumulation
of these. In this manner it may be arrested and re-
established until the fluid is finally exhausted of the fermentable principle. It is well known, for example, that alcohol
will arrest vinous fermentation, that acetic acid will arrest
acetic fermentation, that butyric acid will arrest butyric
fermentation, and that lactic acid will arrest lactic fermentation ; in fact, there is strong evidence for the belief that
this action of bacterial products upon the organisms which
respectively produce them is quite general in its operations
and that its bearing on the history of infectious diseases
and the causes of immunity is of great interest, and entitled
to critical investigation.

That pathogenic bacteria are influenced in their activities
by this law has long been known, and frequently com-
mented upon. For example, Dr. Klein in discussing this
matter says : "One of the most interesting facts observed
in the growth of septic micro-organisms is this; that the

products of the decomposition, started and maintained by them, have a most detrimental influence upon themselves, inhibiting their power of multiplication ; in fact, after a certain amount of these products have accumulated, the organisms become arrested in their growth, and finally, may be altogether killed.''

This was written by Dr. Klein as long ago as 1885; since then much valuable information, upon this and other subjects relating to the history and behavior of bacteria, has been obtained. The facts, as now viewed, lead us to believe that the products arrest fermentation by inhibiting rather than by destroying bacteria; this is certainly what occurs in their action upon ferment organisms; they are not killed, only inhibited by the products, and when these are removed, the organisms are again ready for work.

Now, it is very important to know by what method these products inhibit bacteria ; such knowledge will be appreciated when we remember that the power of inhibiting bacteria, which their products have, is not only intimately, but causatively connected with the fact that acute infectious diseases are self-limited in their duration, and that the accumulation of these products within the body of the infected individual, must in large measure cause the final arrest of the disease. It is therefore strange, but nevertheless true, that none of the theories of infection are capable of giving a rational explanation of this important law; they recognize the law and its importance, but have no explanation to offer of its philosophy. This, however, is not true of the physical theory; it has an explanation to offer, which, it is hoped will prove satisfactory to the reader. Whether or not this much desired result is obtained, the theory is, at least, consistent with itself; is successful in its explanation of the phenomena, and is rational in its methods, inasmuch as these methods are based upon known laws of energy.

If the products of bacterial action,—such, for example, as the poisonous albumins which have been mentioned,—are regarded as interfering bodies, i. e., bodies whose molecular motions give rise to ether vibrations which recur in periods of time that interfere with those produced by the bacterium, and if the bacterium owes its energy to the etherial wave motions which the vibration of its molecules produce, as claimed by the physical theory, then any substance that produces etherial waves which interfere with those of the bacterium, will tend to inhibit its energy.

It will, I think, be conceded that this argument is sound, and the explanation which it offers is sufficient, provided the premises upon which they rest are true. The evidence which supports a part of the premises,—that relating to molecular vibrations in periodic time, and the energy of etherial wave motion,—has already been presented and will not again be referred to. But there is another portion which requires further notice, and, no doubt, the reader has already framed in his mind the inquiry,—upon what evidence is it assumed that these products are interfering bodies? and, if they are such, how does it happen that each product is restricted in its inhibitory action to that bacterium which produced it? To answer these questions clearly, it is better that a restatement be made of certain matters that have already been mentioned, viz.:

1.   A bacterium, or ferment, can decompose, or shake in sunder, those substances only whose molecular vibrations recur in the same periods of time as those of the bacterium.

2.   A substance to be acted upon,—fermented or decomposed,—must be less firmly fixed in its chemical union than the acting cause, ferment or bacterium.

3. The products of bacterial action, or of ferment-bacteria, vary with the nature of the decomposable or fermentable substance and the bacterium or ferment involved; that is, every infectious bacterium and every ferment, has its specific products.

When the micro-organism of vinous fermentation—the yeast cell—through the vibration of its etherial waves, drives the sugar molecules of the grape juice beyond their chemical bonds and thus disrupts the sugar, its molecules are left in a nascent condition; eager to immediately combine into simpler and more stable substances. But it is evident that no new combination can form under these influences, for reasons already given, unless it has a molecular structure that can not be influenced by the wave vibrations of the yeast; the more perfectly a substance coincides in its periods with the yeast, the more difficult will be its formation. The less frequently their periods coincide, the more easily can the new substances form, but in the same degree that the wave vibrations of the yeast differ from those of the new substance, will the waves of the latter interfere with those of the former, and when the new substance accumulates in sufficient amount, it will inhibit the action of the yeast. Thus the products of vinous fermentation—alcohol, carbon dioxide, etc.—must be regarded as interfering bodies whose wave motions antagonize, or inhibit those of the yeast, and when they accumulate will, through their interference, arrest the decomposition of the sugar. For the same reason the products of acetic, butyric and lactic acid fermentations must be regarded as interfering bodies which are capable of bringing to rest their respective fermentations.

If the remarkable similarity which exists between the phenomena of fermentation and infection depends on a like similarity of causation, and the same physical principles which enable ferment-bacteria to disrupt and convert fermentable substances into inhibitory products, operate in giving pathogenic bacteria the power of disrupting albuminoids, then the products which will arise from this action, for like reasons, must have the power of inhibiting the action of pathogenic bacteria. And, as the nature of infectious products will vary with that of the bacterium and albuminoid concerned (as ferment products vary with the ferment and fermentable substance concerned), it is apparent why the products of a ferment or a pathogenic bacterium will inhibit that ferment or bacterium only which produce them; or, at least, why they are limited in their inhibitory action to substances having opposing wave motions.

The important law that acute infectious diseases are self-limited in their duration—will pursue a typical course and be terminated by the action of their own laws—is in a large measure based upon the inhibitory power which the resulting products,—the poisonous albumins, whether termed toxines, tox-albumins or albumoses—exert upon the action of the respective bacteria. But, as these products can act only as interfering bodies during the time they are present in the body of the infected individual, and as they are chemical substances, and, like other similar substances, are soon eliminated, other factors than those furnished by bacterial products must be present in order that the law of self-limitation may hold; whilst the inhibition which is exerted by the products of ferment bacteria are sufficient to permanently arrest fermentation, inasmuch as the product remains permanently in the fermented fluids, such is not the case with pathogenic bacteria, whose products are not permanently retained in the animal body; consequently other factors are needed to explain the resulting phenomena.

The retention theory of Chauvease, it is true, is not in full accord with these statements. The advocates of this theory believe the formation of tox-albumins is the final act, and these bodies the final cause of acquired immunity. The duration of immunity would, from this standpoint, correspond to the length of time the tox-albumins are retained in the body; when immunity is permanent they require us to believe that these are permanently retained, and, that immunity from several infectious diseases results from a retention within the body of several forms of poisonous albumins. The improbability of this hypothesis, its antagonism to physiological methods, and other objections which will be given when this theory is more fully discussed, cause us to dissent from its conclusions and seek the cause of immunity in other directions. Now, the feature of self-limitation of infectious diseases is, we believe, intimately associated with that of acquired immunity; in fact, it is immunization that limits the duration of these diseases. The immunity may be of brief or long duration, but it must exist before the disease can be terminated. We will, therefore, reserve our explanation of this characteristic until the subject of acquired immunity is discussed.

CHAPTER V.

NATURAL IMMUNITY A RESULT OF PROGRESSIVE DEVEL-
OPMENT OF MOLECULAR STRUCTURE, FROM PRIMORDIAL
PROTOPLASM TO HIGHLY SPECIALIZED CELLS, BY NAT-
URAL SELECTION, ADAPTATION AND INHERITANCE.

The remarkable similarity in the phenomena of fermentation and infection leads us to believe that they have a like
similarity, if not identity of cause, and that this rests in
molecular physics. While it is not our purpose to again
discuss at length the analogies of fermentation and infection, or the rationale of their action, it will, at times, be
necessary, as in the present instance, to restate some things
that others may be better explained or illustrated.

The following are well attested facts, viz.:

1. Both fermentation and infection may be induced by
one-celled micro-organisms.

2. These micro-organisms have definite and distinctive
energy.

3. The micro-organisms of fermentation require contact
with fermentable subtances to induce fermentation, and, as
they furnish nothing from their own substance to the ferment products, they must act as physical and not as chemical agents.

4. The products of both fermentation and infection are
inhibitory bodies, and their accumulation will arrest that
process of which they are the result.

If now the teachings of analogy are to be relied upon,
we must admit that pathogenic bacteria cause patho-

genesis or infection, by the dynamic energy resulting from their molecular structure. Certain other important conclusions will be reached by reasoning inductively from these premises, e. g., if pathogenesis is a result of two factors, viz., a bacterium and an albuminoid substance of the animal body whose molecular vibrations coincide in time and periods, then it is apparent that bacteria with molecular vibrations which do not coincide with any of the albuminoids, say of man's body, would be non-pathogenic or harmless to him, but, at the same time, this bacterium may find susceptible albuminoids in the bodies of other animals, and would, therefore, be virulent to them. It follows, then, .that bacteria are pathogenic to those animals only, whose bodies contain albuminoids which vibrate in unison with such bacteria; and, on the other hand, animals whose albuminoids do not have the requisite molecular vibration, are immune from this bacterium. We would, therefore, find as a result of our inductive reasoning that (1) out of the innumerable forms of bacteria some would be pathogenic and others harmless; and (2), that bacteria can be pathogenic to some animals and harmless to others. While these results are logical conclusions from the premises, they are of far more importance; they are matters of fact.

Observation and experience teach that we are surrounded by countless numbers of bacteria; that the food we eat, the water we drink and the air we breathe are more or less contaminated by microbes, and if it were not for the fact that many of these are harmless, and that we have natural immunity from others that are virulent, man would soon cease to exist. Therefore the law of natural immunity, to which man owes, in large measure, his protection from these bacteria, is of great practical importance, aside from its scientific interest, with which our inquiry is concerned especially; it is a law that influences all living things, both animal and vegetable, from the simplest cell to the complex organized structure, from moner to man; and it is to the operation of this law that we must ascribe racial, and, in some instances, individual immunity from infectious diseases; for example, the immunity of the negro race from yellow fever. And yet, strange to say, none of the theories of immunity have even attempted to give it a philosophical explanation. I do not wish to be understood as saying that explanations have not been given, for the phagocytic and humoral theories both have done this; the first by stating that the amœboid cells of the animal body, under certain conditions, destroy virulent bacteria, and the other that certain substances of the blood serum of immune animals neutralize the poisonous products of bacteria. What I do say is, (1) that neither of these theories is sufficiently broad in its scope to take in all the phenomena of immunity, and (2) no explanation is given how the phagocytes of an animal naturally immune from a virulent bacterium obtained their power of destroying this bacterium. (Animals not immune are infected by it.) Or by what process "defensive proteids" are formed in naturally immune animals. A simple statement does not satisfy the demands of science, and we hold, that until the intimate nature of these processes is known, they do not offer a philosophical explanation of natural immunity.

A study of this natural law is approached from two directions. The first comprises a study of pathogenic bacteria, and the second a study of what is termed the "natural resistance of the tissues." A study of virulent bacteria reveals the fact that virulence is a variable quantity; bacteria that are virulent to some animals are not so to others even of the same class; for illustration, the bacilli of glanders is quite virulent to field mice, and harmless to white mice, and the bacilli of anthrax, which are virulent to white mice, are not so to white rats; it would seem, therefore, that this quality of bacteria is active only in susceptible animals. The natural resistance of the tissues, whatever this means, is a favorite expression used to explain (?) natural immunity, as shown by the following: " The problem of natural immunity requires two factors: the first shall be aggressive in its habits and would have the power to invade the body and elaborate therein its poisonous products which produce the specific types of disease. The other factor would be that condition of the tissues of the body whereby they are able to resist the invaders, or destroy them, or prevent the formation of their poisonous products should they succeed in gaining entrance into the body. The first factor is the pathogenic bacterium, the second factor in the natural resisting power of the tissues of the body."* While we do not challenge this statement as a question of fact, we must insist that the explanation of natural immunity which it offers, is unscientific; a mere statement of one unknown process—the vital resistance of the tissues—does not explain the philosophy of another— natural immunity.

"In a series of articles published in *The Lancet*, in 1888, upon the pathology of infectious and infective diseases, the well known pathologist, Joseph Coates, of London, states his conclusions as follows:† 'We have seen that, in the case of a large number of diseases of this class, inheritance,

---

*Frankel's Bacteriology.

†J. Wellington Byers, Reference Handbook of Medical Sciences, Art. "Influence of Race and Nationality upon Disease."

whether we take it more broadly in the race or more particularly in the family, has an undoubted, and frequently a
very great influence on the susceptibility to infection. This
varying degree of susceptibility exists in the case of diseases
which are demonstrably due to the action of micro-organisms, and we are driven to the conclusion that a micro-
organism which is pathogenic to the individuals of one
race to a high degree, is non-pathogenic, or nearly so, in
the individuals of another race.   Recurring to our remarks
on the general principles of inheritance, it seems necessary
to refer this difference in susceptibility to fine differences in
the structure and activity of the tissues.   We have seen
that the differences in the races depend on variations in the
details of their tissues, such as might singly seem to be of
comparatively little moment. The differences in individuals
of the same race are still more minute, and depend on still
finer variations in the details of the tissues.   Coming to
undoubted cases of inheritance of morbid conditions, we
saw by illustration that it was variation in the details of
structure and function which are the subject of inheritance.
In icthyosis it is the structure and mode of growth of the
epidermic cells; in hæmophilia it is the structure, presumably, of the blood-vessels; in Daltonism the finer details
and structure and function of the retina.

"'In all these cases it is the structure and activity of the
finer elements of the tissues concerned, that are at fault.
When we find that the varying susceptibility to infectious
and infective diseases is also related to inheritance, then we
must, I think, relate it to the same kind of variations as
those which we have found to be the subject of inheritance,
both to normal and abnormal structures.   It is, again, the
structure and vital activities of the elements of the tissues
with which he have to do.   If, as we have seen, the negro

race differs very markedly in its susceptibility to infectious diseases from the European races, then we are led to believe that this depends on fine differences in the structure and activity of the tissues, such as determine the characters of the race. It is the living activities of the tissues with which we have to deal, and it is peculiarities in these, determined by inheritance, which, I believe, constitute the differences.' "

What are the fine differences in the structure and activity of the tissue to which Professor Coates refers, but fails to explain? and how do these differences in the structure and activities of the tissues give immunity?

Before answering these questions we desire to state our belief that the chief difficulties which stand in the way of a correct knowledge of immunity, are the obscure vitalistic conceptions which are entertained regarding the physical forces of the body; it is customary to speak of these as "vital forces" in such manner as to lead one to believe that they differ from, and are not correlated to the physical and chemical forces outside the body; that in "vitality" and "vital force" we have an unknown entity which presides over living matter. Therefore when these terms are used in this sense, it is to express an unknown process; for example, if we say that the "activities of the tissues" to which Professor Coates justly attributes much importance in resisting infection, result from vital force as usually understood, then immunity must necessarily be an unknown process caused by an unknowable force and is, therefore, placed beyond the limit of human investigation. It is not, however, in this light that we propose to view it. Vital force we regard as transformed physical force, which is subject to the same laws of transmutation and correlation that govern physical force found elsewhere. It has already been stated, that all living matter, both animal and vegetable, is made up of cells, and these are composed of molecules which are grouped and arranged in definite order. When, now, we regard the tissues of the body as orderly aggregations of specialized cells, and further, regard the cells as molecular machines, and cell structure as depending upon molecular construction and groupings, we have considerably simplified the subject of inquiry, tissue structure, and placed it in a position where further investigation will reveal the fine differences and activities referred to.

In looking back, with our mind's eye through the vista of time, a period in the history of creation will be reached when cells were first formed, and it is not an unreasonable belief, that anterior to this period there existed a structureless, formless protoplasm from which cells were derived. Now the laws of descent and heredity apply with the same force to cell-life that they do to organisms of complex cell-life, and the development of unicellular organisms occurs along the same lines and is governed by the same laws which control the development of multicellular organisms; therefore, in the early history of cell-life, favored by the peculiar environmental conditions which prevailed at that time, a multiplicity and diversity of cells developed from primordial protoplasm, and the differences manifested by the innumerable varieties of cells thus created would be found in molecular structure. Now, molecular structure not only gives to cells their special energies, but it also determines the lines of development which species and variety pursue — and, as well the extent of development which they attain. Therefore, in consequence of molecular structure, some cells would find their fixed limit of development in unicellular organisms, others would attain a higher limit, and develop into multicellular organisms, while still other varieties, because of their molecular structure, would develop into highly specialized cells of complex multicellular beings, such as man. But, as the progressive development of cell structure is determined by the same environmental agencies, and governed by the laws of descent which control the evolution of higher organisms, it becomes necessary that this influence be inquired into. Going back, then, to the period of time when cells were first evolved in infinite varieties from primordial protoplasm, we will attempt a description of the intense and incessant warfare which was then waged by the cellular and amorphous structures against their environment, and by the different varieties against each other.

First, the newly evolved cells were environed by certain conditions of heat, moisture, food supply, light, electricity and climatic influences, to which their very existence was conditioned. Consequently those cells which were in most perfect harmony with their environmental surroundings are the ones which survive the "struggle for life," while those not in harmony would perish.

Second, as cell energy,—the power of doing cell-work,— is a result of cell-structure, and this structure is again a result of molecular grouping within the cell-body, and, as molecules vibrate in definite and distinctive periods of time and thus produce in the surrounding ether, wave-motions of equal periods, it follows that among the infinite varieties of newly created cells there must have resulted a great conflict; the wave-motions of one variety would interfere with, destroy or drive in sunder the molecular combinations of other varieties, the molecules of which would again recombine to form others which would again be destroyed by, or would destroy other cells, and thus a war of cells would continue on and on until there would finally result from the

action and reaction an adjustment of their molecular motions. Now, an adjustment of the molecular forces of the
protoplasm of the many varieties of cells which were en-
gaged in this warfare means that these cells are no longer
warring bodies, their antagonisms have been settled, their
forces adjusted, and henceforth they can live together in
peace and harmony. The world, however, is large and
comprises many climates which have developed cells and
cellular growths in harmony with different environmental
conditions; for example, every country has its peculiar
bacteria, its flora and fauna; consequently, cells that may
be innocuous to other cells which have been subject to the
same environmental conditions, may be, and, as a matter of
fact, are virulent to cells which have developed in another
country under other conditions. Thus bacteria, plants and
many animals, when carried to a foreign country may survive for a time, but sooner or later they are sure to succumb
and disappear under the changed conditions of life.

Recurring again to the primitive state of cell-evolution,
we will endeavor to trace the influence which heredity has
upon the progressive development of cell growths and cellular structures.

From what has been said, it appears that those cells which
are best in harmony with their environment, and best able
to resist injurious influences of other cells, are the ones
which live and transmit their molecular structures to their
progeny, while other cells, which do not possess these qualities, perish in the struggle for life. In the statement that
the fittest survive and transmit to their progeny those
structural characteristics which enabled them to make a
successful race, the term "molecular structure" must be considered in its fullest meaning; it not only includes molecular groupings within the cell, but also the motions in

periodic time which are inseparably associated with such grouping, for it is the molecular vibrations of the cell and the resulting ethereal wave-motions, that gives the cell its energy, its ability to select an appropriate aliment, and its power of attack and defense.   Therefore, when we say that it is the molecular structure of the cell that is transmitted by inheritance, it must not be overlooked that the qualities which go with molecular structure are also transmitted. Inherited characters would then mean the molecular groupings in the cell, and the associated motions which the cell has acquired in its struggle against its environment, and its warfare with other cells.   Now, one of the qualities which the successful cell has acquired and transmits to its successors, is an ability to successfully resist the ethereal wave vibrations that are produced by other, antagonistic, cells; in other words, cells which have inherited these qualities are immune from attacks made upon them by other cells.

No correct opinion of cell-life or the life of cellular organisms can be had; there can be no correct conception of the causes of natural immunity, whether possessed by a bacterium or by those cells of which man's body is composed, that is not based upon the evolution of cell-life from the simple to the complex, and from the undifferentiated to the differentiated cell-structure, as a result of those laws of inheritance known as natural selection, heredity, adaptation and environment.   If our physical theory is true, it is in this manner, and in conformity with these laws that man has acquired his natural immunity; the molecular structure of the cells of his body, with their equivalent molecular vibrations, represent certain adjustments of molecular energy which were gradually acquired by the ancient ancestors of these cells through wars and battles the like of which has never occurred in the history of man.   By "natural selection" is meant that law of nature whereby she selects from a large number of organisms those varieties and species which are in best harmony with their environment. Heredity we understand to be that natural law by which organisms transmit through the race those qualities, possessed by varieties and species, which determine their selection by nature. Adaptation we understand to be that law which enables organisms during their generation to adjust themselves to their environment; qualities thus acquired may be, to a limited extent, transmitted through inheritance. Environment comprises those natural causes, e. g., climate, food-supply, temperature, moisture, etc., which determine from what species or varieties nature will select for perpetuating the race.

It is well known that variations occur within the limits of every race, species or class of organisms. No two individuals are alike, differences of size, vigor, adaptability, resisting power, or some other features, are to be found in every family, order or group of organisms; some varieties of the same class are better able to make the race of life than others because they are in better harmony with their environment, and better able to resist the attacks of other organisms, consequently, these are the ones' which nature selects for race preservation and progressive development. The continued selection by nature of the fittest variations of successive generations, in long periods of time, is, perhaps, the most important factor in the progressive development of the organism, but at the same time, this development is conditioned and limited by the molecular arrangement of its primordial protoplasm. It is in the adjustments of molecular structure, and the resulting adjustments of cell-forces which the primitive cells acquired and transmitted through the favored varieties from which nature selected, in forwarding the progressive development of the organism, and the preservation of these qualities in the germ-cells, through which the race characteristics are perpetuated—that, we believe, the principal cause of the natural law of racial and individual immunity is to be found.

"The structure and activity of the finer elements of the body," and the "variation in the details of the tissues," to which Professor Coates ascribes natural immunity from infectious bacteria, and the cause and inheritance of certain morbid conditions of the body are, from our standpoint, comprised in the condition which we have termed molecular structure; it is this that determines the form and energy of the fluids and solids of the body and, as before stated, molecular structure is, first, a result of molecular warfare and, second, a quality of inheritance. Now, the intricacies and fixedness of molecular groupings vary directly, as do the tissues from which these are formed; the fluids of the body, its plasma and juices, which are largely constructed of albuminoid molecules, are relatively simple in molecular structure, while the solids of the body, which are composed of highly specialized cells, are relatively complex in molecular structure. The simpler structures found in the fluids, for example, the unstable albuminoids, of many isomeric forms, are far more vulnerable to the dynamic energy of bacteria or other agencies than the fixed cells, and are, therefore, those most intimately concerned in the development of infection and the production of immunity.

When the statement, that natural immunity depends upon the natural resisting power of the tissues to pathogenic bacteria, is examined in the light furnished by this physical explanation, it ceases to be jargon and becomes, at once, an intelligible hypothesis in which the mind has no trouble in clearly tracing the relationship and mode of

action between the cause assigned, and the effect claimed
by the hypothesis. Natural immunity is, however, not
alone secured by natural selection in the manner above
described. There can be no reasonable doubt that the constant presence of, or even frequent visitations from virulent
infectious diseases, will first destroy those who are most
susceptible to its influence, and, in the course of time but
few, if any out of the entire population, will remain who
are not naturally immune from the infection; and as natural
qualities are transmissible through inheritance, a time
would arrive, sooner or later, when a race of people placed
under such influences, would become naturally immune
from an infection that would be virulent to other races of
people who had not been similarly placed.

It is attested by physicians, whose residence has given
them excellent opportunities of observation, that persons
who are constantly subjected to endemic, or who are frequently exposed to epidemic influences, will acquire a
greater or less degree of immunity from such endemic or
epidemic diseases. Immunity acquired in this way is,
however, a different thing from that which is secured by
the natural law which we have been discussing; it is limited in its action to the individual, and never transmitted by
him to his children; whereas natural immunity is a racial
characteristic which protects not alone the individual, but
also gives immunity to his posterity through indefinite generations. The first is acquired and not transmissible, the
other is inherited and is transmissible. Immunity may
also be acquired by other methods, but, however this is
induced, it never extends beyond the life of the individual;
but often, unfortunately, is terminated long before this time.

The subject of acquired immunity is one of great interest,
but we must reserve its discussion for the next chapter of

this essay. Whether such qualities are inheritable is a subject of dispute, the latest and more advanced views, based
upon the investigations of Weisman, declare against the
transmission of acquired characters, but this opinion, we
believe, does not hold true of the lower forms of life, especially those forms,—the bacteria,—with which our inquiry
is concerned, for it has been conclusively established by
experimental investigations, that bacteria, and, perhaps,
other low forms of living organisms, will transmit through
many generations certain qualities which .they have
acquired by artificial means; for example, from exposure to
excessive heat, from the action of chemical agencies, or
from growing such organisms in unsuitable food media.
This process, changing the biologic habits of bacteria, is
known as "attenuation" and is of great interest and practical importance; it especially concerns our inquiry, because
of the direct bearing it is believed to have upon man's
acquired immunity against infectious diseases, and will be
further referred to under this head.

The reason why bacteria and other low types of living
organisms, are capable of transmitting through the race,
qualities which they have acquired, while complex multicellular organisms of the higher types of organization do
not possess this power, is no doubt found in the fact that
there are no visible structural differentiations in the simpler
cells; such cells have no distinguishing parts upon which
devolve a separate performance of physiological labor; on
the contrary, there is a homogeneity of structure and physiology in all portions of such cells; the same physiological
qualities, of breathing, eating, digesting, assimilating, moving, and of reproduction, which the cell as a whole possesses, may also be performed in full by any of its divisions.
But in the specialized cells composing the bodies of higher

organisms, a different state is found to exist; certain cells are set apart to perform special work, thus one class of cells perform this, another class of cells perform that, and other classes perform different functions of the organism. Now, in this dividing out among the different cells of the body the physiological labor of the organism, giving to each class of cells that portion of this labor to which its structure and organization is best adapted to perform, and for which, in fact, it was essentially evolved, it is found that the physological function of reproduction is likewise entrusted to a class called germ-cells, whose single duty is to perpetuate the race; they belong to the race and transmit its characteristics, and not those acquired by the individual. The somatic or body cells, on the other hand, belong to the individual, and all changes which are produced (through adaptation) in these cells or in the tissues of the individual's body, terminate with his life. Thus a sharp line of division is drawn between the germ-cells and the somatic cells; the first belong to the race, the second to the individual; changes induced in the germ-cells will be transmitted through the race, while changes induced in the somatic-cells continue only during the life of the individual. Natural immunity, then, is a characteristic of the germ-cells and is transmitted by inheritance, while acquired immunity is a quality of the somatic cells and, consequently, is not transmissible, but terminates with the life of the individual.

The kind and degree of immunity which nature secures to cells and cellular organisms, will therefore vary with the conditions of invironment under which the cells and cellular organisms have developed; other factors are the conjugation of different races and varieties, and the perpetuation of acquired characters which are transmitted by heredity.

Man's natural immunity from the vast swarms of bac-

teria which surround and envelop him, is then to be ascribed to the fact that the cells and cell molecules of his
body are invulnerable to the disruptive power of bacteria
whose molecular vibrations cannot shake apart and convert
into poisonous albumins any of the albuminoid molecules
of his body.   But we must not overlook the fact that the
law of natural immunity is not absolute in its operations.
While it is doubtless of wide application, still, there are
many exceptions to this, as there are to that of progressive
development by natural selection and inheritance in other
departments of natural history.   The exceptions in this
case comprise what is termed "susceptibility," which is the
direct opposite of immunity.   Neither of these conditions
of the body—susceptibility or immunity—is absolute in degree or duration; for example, that degree of immunity
which will protect an individual from infectious bacteria
when they are introduced into the body by ordinary methods, and in comparatively few numbers, will not protect him
from these microbes when they are introduced by unusual
methods, say directly into his blood, or in large numbers at
one time; under such conditions the immunity of the body
may be overcome, and its infection accomplished.   The duration of immunity, like that of degree, is also subject to variations; while these changes are observed more frequently
in acquired immunity than in that form which is secured
by natural law; yet even this may be destroyed by artificial
means; this is indicated in the cases reported by H. Leo.
"The bacilli of glanders are, as we know, particularly virulent in their action on field mice, while white mice, the animals most frequently experimented upon, are quite insusceptible to them.   H. Leo has, nevertheless, recently succeeded in making them susceptible to glanders by feeding
them for a length of time on phloridzine, thereby putting

them artificially into a diabetic condition and saturating their tissues with secreted sugar.*

Immunity and susceptibility must then be regarded as relative terms which imply that these conditions of the body produce immunity from, or susceptibility to the dynamic energy of a pathogenic bacterium, and that this immunity and susceptibility are qualities that are variable in duration and degree. We have shown already how cells, cellular organism and amorphous protoplasm acquire, by molecular dynamics and progressive development through natural selection and inheritance, immunity from other similar bodies; and as susceptibility is the opposite of immunity, it follows that the conditions of environment under which susceptible organism developed, were quite different from those of immune organisms, but as environment comprises climate, food-supply, moisture, atmospheric and telluric conditions, etc., and as these are influenced by geographical localities and varied, perhaps, in geological time, it is seen that tracing out these influences would be an endless if not impossible undertaking. We must, therefore, be satisfied with the statement that the same principles of matter and motion, and of progressive development, are the active agencies in the production of both conditions; and that the difference in results is doubtless caused, principally, by differences of environment. Organism which have developed in one climate or under given environmental conditions, are susceptible to bacteria from which organisms developed in other countries or under other conditions, are immune.

Now, as different races of men have developed under different conditions of environment, it follows, from the laws of natural selection and descent, that corresponding differ-

*Frankel's Bacteriology.

ences exist in the molecular structure of their albuminoids, and these carry with them differences in susceptibility to bacterial influences.   Man's migratory habits, his ability, in many cases to withstand environmental influences, and the intermingling of races, it is true, greatly modify the action of these laws, so far as they relate to his present state of development, but it is a reasonable presumption, that in the earlier stages of his development, or those of his tissue cells, such favorable conditions did not prevail, and consequently we would expect to find, and as a matter of fact there does exist between races of men, a marked difference in susceptibility to infection.

The underlying principle, the *point d'appui* of the physical theory of immunity and infection, is grounded on the accepted teachings of physics, chemistry and biology, while the theory in its full development is the outcome of what is believed to be legitimate deductions from these teachings.   For example, it is believed and taught that atoms and molecules vibrate in definite and unvarying periods of time; that atomic and molecular vibrations produce wave-motions of equal periods in the surrounding ether; that ethereal wave-motions are subject to the "law of interference;" that force, the efficient cause of all physical phenomena, whether manifested as physical action outside the body, or as physiological action inside the body, is inseparably connected with the motion of atoms, molecules or of mass;* that atomic and molecular energy is the result of

*The force referred to is physical force, and its domain of action is confined to physical phenomena within or without the body; it is not claimed nor believed, that this is the efficient cause of mental phenomena when these are considered aside from the physical causes which are required to develop them.   Such phenomena do not belong to the domain of physical science, and are not explicable by any

wave-motions produced by atomic and molecular vibrations in periodic time; and for the same reason, and in the same degree that molecular wave-motions are influenced by interfering waves, will the energy which they produce vary in its manifestation, or degree, with that of its cause.

The wave vibrations produced by a molecule are, therefore, regarded as the sum of its atomic vibrations, and the result of an adjustment by interference, of atomic wave-motions.

Therefore, it is assumed that the ether waves or energy of a molecule, or molecular compound, will vary with molecular structure, and, when such compounds are formed into organized or living tissue, the energy which results from its molecular movements constitute "vital force"

---

physical hypotheses. The distinction between mental and physical phenomena is thus forcibly stated by Professor Huxley: "Nobody, I imagine, will credit me with a desire to limit the empire of physical science; but I really feel bound to confess that a great many very familiar and, at the same time, extremely important phenomena, lie quite beyond its legitimate limits. I cannot conceive, for example, how the phenomena of consciousness, as such, and apart from the physical process by which they are called into existence, are to be brought within the bounds of physical science. Take the simplest possible example, the feeling of redness. Physical science tells us that it commonly arises as a consequence of molecular changes propagated from the eye to a certain part of the substance of the brain, when vibrations of the luminiferous ether of a certain character fall upon the retina. Let us suppose the process of physical analysis pushed so far that one could view the last link of this chain of molecules, watch their movements as though they were billiard balls, weigh them, measure them, and know all that is physically knowable about them. Well, even in that case we should be just as far from being able to include the resulting phenomena of consciousness, the feeling of redness, within the bounds of physical science, as we are at present. It would remain as unlike the phenomena we know under the names of matter and motion as it is now."

which, again, varies with molecular structure; thus the "vital force" or energy of a brain-cell is quite different in its results from that of a liver-cell, and this difference, it is believed, is the result of molecular structure.

When, now, we regard the progressive development of cells and cellular organisms, thus constructed and controlled, as taking place under the rule of "natural selection" and "heredity," I think it must be admitted that the conclusions of our physical theory are the outgrowth of fair and legitimate reasoning from accepted premises.

These views may, no doubt, appear quite startling to many persons, as do all declarations which seriously disturb the usual order of prevailing opinions, but they certainly should not be discarded for these reasons alone. If they are true they should be accepted, if not true their fallacy should be exposed.

Unfortunately medicine has gathered, during its long career, much useless and unprofitable lumber, such as old sayings like "vital force," and old theories like "vital action," which are seriously retarding its progress into the galaxy of exact sciences; they should be thrown overboard, or gotten out of the way and a new start should be taken with clearer and more scientific ideas. The following quotations from men no less eminent than Foster, in Physiology; Huxley, in Biology; J. Clerk Maxwell, in Physics; and Thomson, in Chemistry, show the trend and scope of scientific thought on these questions. It is needless to say that the opinions of such men, in matters relating to their special departments of science, are entitled to great respect:

Professor Michael Foster, M. D., says: "We have, in speaking of protoplasm, used the words 'construction,' 'composition,' 'decomposition,' and the like, as if protoplasm were a chemical substance. And it is a chemical substance, in the sense that it arises out of the union or coincidence of certain factors which can be resolved into what the chemists call 'elements,' and can be at any time, by applying appropriate means, broken up into the same factors, and indeed into chemical elements.

"This is not the place to enter into a discussion of the nature of the so-called chemical substances, or, what is the same thing, a discussion concerning the nature of mat- . ter; but we may venture to assert that the more these molecular problems of physiology, with which we are now dealing, are studied, the stronger becomes the conviction that the consideration of what we call 'structure' and 'composition' must, in harmony with the modern teachings of physics, be approached under the dominant conception of modes of motion.

"The physicists have been led to consider the qualities of things as expressions of internal movements; even more imperative·does it seem to us that the biologist should regard the qualities (including structure and composition) of protoplasm as in like manner the expression of internal movements.

" He may speak of protoplasm as a complex substance, but he must strive to realize that what he means by that is a complex whirl, an intricate dance, of which what he calls chemical composition, histological structure, and gross configuration, are, so to speak, the figures; to him the renewal · of the protoplasm is but a continuance of the dance, its functions and actions the transference of figures.   . . . . And the conception which we are urging now is one which carries an analogous idea into the study of all the molecular phenomena of the body.

" We must not pursue the subject any further here, but we felt it necessary to introduce the caution concerning the word 'substance;' we may repeat the assertion that it seems to us necessary, for the satisfactory study of the problems on which we have been dwelling for the last few pages, to keep clearly before the conception that the phenomena in question are the result, not of properties of kinds of matter, in the vulgar sense of these words, but of kinds of motion.''*

Professor Huxley says: ''The broad distinctions, which, as a matter of fact, exist between every known form of living substance and every other component of the material world, justify the separation of the biological sciences from all others.   But it must not be supposed that the difference between the living and not living matter are such as to justify the assumption that the forces at work in one are different from those to be met with in the others.   Considered apart from the phenomena of consciousness, the phenomena of life are dependent upon the working of the same physical and chemical forces as those which are active in the rest of the world.

''It may be convenient to use the terms 'vitality' and 'vital force,' to denote the causes of certain great groups of natural operations, as we employ the names of 'electricity' and 'electrical force' to denote others; but it ceases to be proper to do so, if such a name implies the absurd assumption that 'electricity' and 'vitality' are entities playing the part of efficient causes of electrical or vital phenomena.

'' A mass of living protoplasm is simply a molecular machine of great complexity, the total results of the working of which, or its vital phenomena, depend on the one hand upon its construction, and on the other to the energy supplied to it; and to speak of 'vitality' as anything but the

*Encyclopænia Britannica, 9th ed., Article Physiology.

name of a series of operations is as if one should talk of the horology of a clock."*

Professor J. Clerk Maxwell says: "Thus molecular science sets us face to face with physiological theories. It forbids the physiologist from imagining that structural details of infinitely small dimensions can furnish an explanation of the infinite variety which exists in the properties and functions of the most minute organisms.

"A microscopic germ is, we know, capable of development into a highly organized animal. Another germ, equally microscopic, becomes, when developed, an animal of a totally different kind. Do all the differences, infinite in number, which distinguish the one animal from the other, arise from some difference in the structure of the germs? Even if we admit this as possible, we shall be called upon by the advocates of pangenesis to admit still greater marvels. For the microscopic germ, according to this theory, is no mere individual, but a representative body, containing members collected from every rank of the long-drawn ramification of the ancestral tree, the number of these members being amply sufficient to furnish not only characteristics of every organ of the body and every habit of the animal from birth to death, but also to furnish a stock of latent gemmules to be passed on in an inactive state, from germ to germ, till at last the ancestral peculiarity which it represents is revived in some remote descendant."†

Julius Thomson, in his introduction to "Thermo-Chemical Investigations," says: "Theoretical chemistry is based upon the molecular theory, according to which all matter is made up of molecules, and these molecules of atoms.

---

*Ibid., Article Biology.
†Ibid., Article Atom.

The physical state of bodies depends upon the arrangement and motions of the molecules, the other physical and chemical properties depend upon the kind and number of atoms in the molecule, upon their arrangement and relative motions."*

*W. R. Nichols: Popular Science Monthly, October, 1883.

CHAPTER V.

ARTIFICIAL IMMUNITY: ACQUIRED BY A SINGLE ATTACK;
BY VIRULENT BACTERIA ; BY ATTENUATED BACTERIA ;
BY THE PRODUCTS OF VIRULENT BACTERIA; BY BLOOD
SERUM OF IMMUNE ANIMALS; BY DEFENSIVE PROTEIDS;
BY UNCLASSIFIED METHODS — ATTENUATION OF BAC-
TERIA: HOW PRODUCED; EFFECT OF IN THE FORMA-
TION OF BACTERIAL PRODUCTS — PATHOGENESIS, THE-
ORIES OF: THE SECRETION OR EXCRETION THEORY ;
THE PHYSICAL THEORY.

Immunity is not always a quality of inheritance; susceptibility of a living organism to a pathogenic bacterium can
be destroyed and the organisms thereby made immune from
the action of this bacterium by means of various artificial
agencies. This condition is termed "acquired" or "artificial" immunity, and, like that termed "natural immunity,"
is variable in its degree and duration; for example, common observation teaches that many of the acute infectious
diseases occur but once in the same individual. This
is notably true of yellow fever, scarlet fever, small-pox,
measles, mumps and whooping-cough; one visitation from
any of these diseases produces a change within the body
of the infected individual that protects him during life
from a second invasion of the same diseases. But a single
attack from other acute infectious diceases does not always
confer such lasting immunity, in fact, there is a marked difference in the degree and duration of protection to an indi-

vidual from a single attack of these diseases; for some it is
for life, for others a period of years, or, it may be, during
the season only, while for others the duration of immunity
is even more brief.

Immunity from an infectious bacterium, or, what is the
same thing, from an infectious disease, can be acquired by
the animal organism in many other ways than the one mentioned, but the degree and duration of acquired immunity
will remain a variable quantity regardless of how this condition is produced; e. g., while the degree of immunity is
sufficient to protect the individual from ordinary exposures
it will not suffice to do this, perhaps, when the bacterium
is unusually virulent, or is introduced into the body through
unusual channels, or in unusually large numbers.   And on
the other hand, the condition of the immune organism, its
resisting power, is an important fact; this will be considered later.

The artificial agents or agencies which are known to immunize susceptible individuals from infective bacteria are
as follows viz:

1.   Infectious bacteria and the products of their action—
toxines and toxalbumins—when these are introduced into
the susceptible organisms, at first in small then in gradually increased amounts.

2.   Attenuated bacteria.

3.   Blood serum of immune animals, and certain proteid
substances contained in the blood serum, or tissue juices of
such animals; and

4.   Other, indifferent substances which, apparently, are
not related to either infectious bacteria or to the products
of their action.

Some of the facts of acquired immunity are so apparent
that they may be regarded as an heritage of common obser-

vation; these, and the phenomena which they produce, have been observed from remote times, and speculation regarding the modus of their action has frequently been made and recorded in the medical literature of the time, but were necessarily empyrical, as the subject cannot be scientifically explained until the causes of infection and the rationale of their action are known. Other phenomena of acquired immunity are less apparent, and our knowledge of these has been obtained only in late years and by experimental work. For reasons which are obvious, the experiments that were made to determine these facts of immunity were performed on lower animals, and in many cases the benefits which have resulted from them are confined to these animals. But this is found to not be the case throughout; in some instances man, and some of these animals are alike susceptible to the same infectious bac' rium, and it has been established that immunization wil. result from the same cause in both; for example, the virus of anthrax, rabies and tetanus is virulent alike to man and many of the lower animals, and immunity from these diseases can be produced by like means in both cases. We think, therefore, that the principles which underlie and control infection and immunity are the same regardless of the animal in which they operate, and that difference of results, which are known to vary with the nature of the animal in which these are manifested, occur from differences in the molecular structure of their tissues, and on this depends their ability of resisting injurious influences.

The causes and philosophy of acquired immunity is, perhaps, the most important subject, in its theoretical and practical aspects that is now occupying medical thought; it is the germ of our inquiry to which our previous investigations have tended. This subject is receiving, at this time, more scientific attention from medical writers and thinkers than ever before in the history of medicine. Its practical value, aside from its scientific interest, must be appreciated when we consider that the principles that determine immunity are they which protect us from the vast swarms of microbes which surround and would destroy us, and also give acute, infectious diseases their self-limited feature, to which fact man is again indebted for his life; were it not for this feature, these diseases would continue until death ended the work.

ACQUIRED IMMUNITY INDUCED BY VIRULENT MICROBES OR THEIR PRODUCTS.

It has been proven by the most irrefutable evidence that pathogenic bacteria are the infectious agents of a large number of infectious diseases; when these micro-organisms are introduced subcutaneously into the bodies of susceptible animals they invariably produce, in such animals, the symptoms and pathological lesions of that disease of which the inoculated bacterium is the undoubted cause. And as one attack from one of these diseases will confer immunity to the animal from other attacks, it is seen that infectious bacteria confer immunity to a previously infected animal from future invasions of this bacterium; immunity against a disease is immunity against the cause of that disease. This view of the subject represents that which was generally accepted until later investigations revealed to us that the products of bacteria, and not the bacteria themselves, are the true disease producing agencies;—not that they are capable of producing all the phenomena of infection when, e. g., sterilized cultures of the bacteria are inoculated, but those only that are toxic. Transmissibility of the disease is, of course, dependent on the bacteria. The effect produced by inoculations of these bodies—toxalbumins—will vary with susceptibility and dose. "The pathogenic dose of a virus varies with the predisposition of the animal to the disease in question; the greater the predisposition to the disease, the less is the quantity required, and conversely, the less the predisposition, the larger is the dose required, until ultimately the point is reached when no amount of virus will produce an effect."—(*W. Watson Cheyne.*) The manner of introducing the virus—toxalbumin—is a matter of no less consequence. If the initial dose is small—non-toxic—and the amount is gradually increased in successive and frequently repeated inoculations, the animal will speedily become refractory to very large—toxic—doses, and what is more remarkable, it also acquires an immunity of greater or less duration, sometimes for life, against the tox-albumin, the bacterium which produced it, and the disease produced by these. To this order of phenomena doubtless belong those which relate to immunity from rabies.—(*Pasteur.*)

Pasteur's inoculations against hydrophobia are now well known, and if the success of this method requires other testimonials than those furnished by Pasteur in his published reports, they can be found in Pasteur Institutes in every civilized country. While hydrophobia is not known to be a microbe disease, and cannot be thus classed until the causative microbe has been found and identified, yet there can be but little doubt that this is its true nature; like these diseases, it is communicable and has a well defined period of incubation. These qualities are of themselves almost sufficient to justify its classification as microbic. Whether the improved technique and other means of further research will discover its infectious cause, there can be no doubt that this, whether a bacterium or its products, is located principally in the spinal cord of the infected individual. This is the substance that Pasteur uses for his inoculations; by submitting the cord to a process of rapid dessication he found that its virulence became rapidly lessened, and in this manner he was enabled to obtain a virus of different strength. Beginning his inoculations with the weakest virus and after a short time using a stronger, and next a stronger still, he found that ultimately the strongest virus could be used without danger; i. e., an animal that is refractory to the strongest virus is immune from rabies. And it has been proven, furthermore, that this immunity is of considerable duration. A certain degree of immunity from snake poison has been |induced by frequent inoculations, beginning with a very small dose of this poison and gradually increasing this until an amount that would be fatal if used at first, can eventually be used without harm; the immunity or tolerance thus induced is not lasting but will pass away in the course of a few months. It has been proven by Professor Erlich, of Berlin, that immunity from two poisonous albuminoids,—enzymes, obtained, one from the castor oil bean, called *ricin*, and the other from the jequirity bean, called *abrin*, can be induced by a similar method of procedure; immunizing substances in both instances are albuminoids, and evidently belong to that class of ferments termed enzymes; in the first case the substance is of animal origin, and in the other of vegetable origin; but, if analogy can be relied upon, they act in both cases by contact with the albuminoids of the body; the toxic effects result, doubtless, from the changes which these ferments cause in the albuminoids.

Recurring to the subject of immunity produced by the products of bacterial action we will instance a few examples of this to illustrate what appears to be a general rule, that the products of a bacterium when introduced into the animal body, at first in small then gradually increasing doses, will immunize it against toxic doses of this poison. The products obtained from sterilized cultures of anthrax bacilli inoculated into a susceptible animal will render it refractory to the virus of anthrax. Hankin has succeeded in extracting from the bodies of animals immune from anthrax an albuminoid substance which possesses immunizing qualities. Behring and Kitasato have extracted and separated from animals, made refractory to tetanus, an albuminoid that is protective against tetanus. By a similar method they have obtained a substance from the blood of animals made immune from diphtheria, and by inoculating this substance into susceptible animals they also became immune from this disease. The tox-albumins of cholera, typhoid fever, chicken cholera, pneumonia of animals, of the vibrio-Metschinkoff, and the staphylococcus pyogenese aurius, either in a purified form or in connection with sterilized cultures, have been proved to possess immunizing qualities and are capable, when properly introduced into susceptible animals, of rendering them refractory to the diseases which these bacteria cause.

IMMUNITY PRODUCED BY INOCULATIONS WITH ATTENUATED FORMS OF VIRULENT BACTERIA.

To Pasteur we are first indebted for our knowledge that microbes may be radically changed in biologic habit by compelling them to exist under conditions of environment which are unfavorable to their normal development; by such means the virulence of bacteria may be modified or destroyed, and virulent bacteria, by this means, can be converted into protective vaccines. This process of weakening the disease-producing qualities of bacteria, however it is produced, is termed attenuation and is a matter of prime importance in its relation to the causation of acquired immunity. In 1880, Pasteur published his memorable work on Chicken Cholera, in which he first announced his discovery that attenuation modifies the virus, i. e., weakens the power of bacteria to produce the products of their functional action. This discovery bids fair to revolutionize preventive medicine.

The following extracts embody these features of Pasteur's work: "There appears among the poultry kept in yards, especially among the fowl and geese, a very destructive, murderous plague, the symptoms of which remotely resemble those observed in genuine cholera of man, for which reason it is called chicken cholera (*choléra des poules*)."

"Perroncito, and after him Pasteur, demonstrated the presence of bacteria in the blood, organs, and excreta of affected animals. Pasteur cultivated them artificially outside of the body, and, being able (in 1880) to produce the disease from the cultures, he furnished the incontestible proof of the causative significance of the micro-organism . . . . . . He perceived that cultures exposed for some length of time (for months) to the influence of the oxygen of the air (*i. e.*, preserved with simple wadding without any other means) lost their virulence more or less, and were no longer injurious to animals. New generations could even be obtained at will, all of which preserved the same attitude. Whenever Pasteur inoculated such material into the breast muscle of chickens, for instance, a mere local inflammation ensued, which generally became rapidly circumscribed and terminated in the expulsion of the altered tissue by suppuration, without any other disturbance . . .

By means of these inoculations at first with a greatly weakened infectious substance ("*le prêmier vaccin*") and subsequently with a much stronger ("*le deuxieme vaccin*"), even highly susceptible animals, such as fowls and pigeons, may be secured against infection."*

More recent investigations have proved that many other infectious bacteria not only lose their infectious power by attenuation, but, at the same time, acquire that of a protective vaccine, which, if introduced subcutaneously into the bodies of susceptible animals, will render them immune from the poison of the virulent bacteria.   The bacillus of anthrax, of hog erysipelas, of mouse septicæmia, of black leg, vibrio Metchiukoff, and many other varieties, can with certainty be changed into protective vaccines by attenuation.   But these are established facts which have become a part of medical history, and it is the causes which produce, and the intimate nature of attenuation, that especially concerns our inquiry.

Accepting, then, the established fact that inoculation with attenuated bacteria will immunize the individual, that is, will often protect him from the virulent form of the same bacteria, we will discuss the means by which pathogenic bacteria are attenuated, the intimate nature and causes of attenuation, and further on will explain how immunity is secured by inoculations with these organisms.

The following extract "from Frankel's Bacteriology" will furnish the answer to our first inquiry:

"In the year 1880, Pasteur surprised the scientific world by the discovery that under certain circumstances the micro-organisms of chicken cholera lose their virulence, to a greater or less extent, without showing any change in their appearance, way of growth, etc.

"Toussaint and Pasteur found that anthrax bacilli could

---

* Frankel's Bacteriology.

be deprived of their virulence in like manner; and the same fact has since been proved with regard to swine ery-sipelas, symptomatic anthrax, the pneumonia bacteria of A. Frankel, and some others.

"This diminution of virulence is the result of two essentially different causes. The one, which we may call the natural cause, has recently been fully elucidated, particularly by Flügge. It is a gradual diminution of infectious power in bacteria which we compel to vegetate for a long time separated from their natural conditions of growth, on our artificial food media, and under the atmospheric conditions of our laboratories. By a gradual adaptation to the altered, saprophytic way of life, or by a progressive selection of such cells as are naturally more capable of adopting this altered way of life, the original capacity for development within a foreign body diminishes more and more. As an outward sign of the change which has taken place, we observe that the culture now shows a much more luxuriant and rapid growth on the lifeless food medium than was at first the case, when the conditions were yet new and strange. Not all the pathogenic bacteria possess this power of 'cutting their coat according to their cloth' and adapting themselves to outward circumstances. Some cling with marked tenacity to their proper character, which they do not leave even when compelled to exist for years outside the body. Others, as the bacilli of glanders, the streptococci of erysipelas, Frankel's pneumonia bacteria, the diphtheria bacilli, etc., lose their virulence very quickly.

"A similar phenomenon was observed in the case of the saprophytic bacteria. Hueppe and his pupils have shown, for instance, that the sour-milk bacillus and the blue-milk bacillus, when bred continually and uninterruptedly on our artificial foods, lose the capacity for effecting the changes from which they take their names, and Hueppe even speaks of this as a loss of virulence in these micro-organisms.

"It lies in the nature of things that this diminution of virulence takes place gradually, proceeding step by step, and not with one great leap. Therefore, we are often able to interrupt the process at a certain stage, or even to retrograde and undo the work already done. The best, and indeed the only, means to this end is to give back to the partially weakened cells their natural conditions, and endeavor to re-accustom them once more to their former way of life.

"In the case of the pathogenic species, we attempt to inoculate them into the animals most susceptible to them. If this fails, we can have a recourse to methods of increasing the natural susceptibility of an animal artificially. Should this prove successful, and the micro-organism once more establish itself on its natural soil, one may reckon with some degree of certainty, on its recovering its former power.

"In direct contrast to the phenomena hitherto discussed, is a second mode of diminishing the virulence of micro-organisms, which leads to the same final results. That which brings about the diminution is, here, not the long continued influence of altered (but not necessarily worse) conditions of life; it is the action, for a short time, of influence directly prejudicial to the bacterial protoplasm.

"In fact, all the means employed to produce artificial diminution of virulence in pathogenic bacteria are such as, if applied in a slightly stronger form, would cause the destruction of the cells, and would kill their contents.

"Thus we breed bacteria in media to which a certain quantity of antiseptic or disinfecting substance has been added, but which just allows the microbes to exist and grow upon them. Such, for instance, as the process of Roux and Chamberland for diminishing the virulence of anthrax bacillus, by cultivating it in a bouillon with the addition of bichromate of potassium (1 to 5000 to 1 to 2000); and that of Toussaint, by mixing about 1% of carbolic acid with blood containing anthrax bacilli.

"In a similar manner—*i. e.*, as a disadvantageous form of nutrition—the organism of certain animals is found to act, namely: those animals which are insusceptible, or but little susceptible, to the particular kind of bacteria which we desire to weaken. Thus the bacilli of swine erysipelas lose their virulence to a certain extent when passed several times through the bodies of rabbits, as has been proved by Pasteur and Kitt.

"Chauveau robbed the anthrax bacilli of their poisonous property, by breeding them under a pressure of eight atmospheres, and Airlong found that "sunlight is capable of weakening these same bacilli, and even their spores."

"By far, the surest and most commonly used procedure is the exposure of the micro-organisms to the influence of high temperatures.

"Toussaint kept blood containing anthrax bacilli for ten minutes at 55° C. The bacilli were by no means killed by the heat, yet they were rendered harmless by it. Pasteur, for his experiment on a large scale, employed a considerable lower degree of heat, but he had not given full particulars of his method, so that no definite judgment can be formed regarding it. We, therefore, owe our thanks to Koch and his coadjutors, who once more approached this question in a methodical, strictly scientific course of experiments, with a view to ascertain the effects of heat in diminishing bacterial virulence by studying the anthrax bacillus—the best known and most suitable species for this purpose.

"Koch, Gaffky and Loeffler found that at 42° C. and 63°

C. a diminution of virulence was perceptible in the cultures.

"They further discovered the important fact that the lower the temperature is by which a diminution takes place, the longer it takes for such diminution, but at the same time the more permanent are the effects.

"Even variations of a fraction of a degree are here important. While anthrax bacilli can be rendered perfectly harmless in nine days with 43° C., it requires twenty-four days if we only employ 42.6° C.; but in this latter case the new quality of the bacteria has become so thoroughly a second nature to them that they cannot throw it aside again. They not only keep it throughout their own life, but they even transmit it to their progeny. In fact, we can breed from them as many generations as we will of fully harmless bacteria.

"If we endeavor to diminish virulence under the influence of higher temperatures more quickly—in a few days—the bacteria regain it with proportional rapidity.

"But their virulence cannot be destroyed at one blow. Before the micro-organisms part with it completely they pass through a number of intermediate stages, each of which, with the amount of virulence still remaining, is sufficient to affect certain animals, the efficacy of the poison remaining longest for those most susceptible to it. Bacilli of twenty days at 42.6° C., for instance, will still kill mice; those of twelve days will kill guinea pigs; those of ten days, rabbits; those of six days, sheep, etc.; and this degree of partially diminished virulence can also be preserved throughout generations of cultures.

"And even if the above statements may not, perhaps, always be borne out in practice with perfect exactitude in all cases, that does not change the incontestably proven scientific fact that bacteria of a high degree of virulence may lose this quality for a longer or shorter time, or even permanently, and to any extent, up to its complete extinction.

"Something similar has been observed in the saprophytic species also. Many pigment bacteria, under the influence of high temperature or of culture media little suited to them, lose the faculty of forming coloring matter, and sometimes do not regain it under normal circumstances till after a considerable lapse of time.

"How is this extremely striking phenomenon to be explained.? What distinguishes bacteria in their normal state from those artificially debilitated?

"Why can the former grow and multiply in the bodies of susceptible animals and the latter can not?

"The circumstance that all the influences which rob the bacteria of this, their (to us) most important capacity are such as are injurious and hostile to them would lead to the supposition that an extensive degeneration of the cell protoplasm took place, which would show itself in other places also. Yet this is found to be the case to a very limited extent only. . The harmless anthrax bacillus has the same appearance and the same form as the normal anthrax; its separate members show the same formation, the contents are as clear as crystal and homogeneous, the rods are motionless, they divide and produce spores as before. On the gelatine plate and in the needle-point culture we observe the same sort of growth; in short, there are no really striking differences."

Of course, these interrogations must be answered, and the subjects involved must be explained before a philosophical explanation, how attenuated microbes induce immunity, can be made. If, now, we accept the results of that experimental work which has led up to and forced the conviction that the products of bacteria are the toxic agents and the

true cause of infection, then it must be that attenuated bacteria produce less amounts of these toxics than virulent
bacteria do; in other words, the process of attenuating virulent bacteria weakens their power of producing the toxic
agents upon which their virulence depends. This view of
the matter, which we have seen is in harmony with that
entertained regarding the true cause of infection, is also
that which is sustained by sound analogies. For illustration, attenuation of ferment-organisms, say of yeast
cells, weakens their power of producing ferment products—
alcohol and carbonic dioxide. Attenuation of pigment bacteria weakens their power of producing coloring matters,
which comprise the products of their functional action, and
analogy teaches that, in like manner, attenuation of pathogenic bacteria weakens their power of producing pathogenic
products. But fortunately we are not compelled to rely
wholly upon analogy for a knowledge of this subject.
Pasteur, from observation and experimental work, has
arrived at the conclusion that attenuated bacteria produce
the same products, only in smaller quantity, that are produced by bacteria not attenuated. This view is that which
we advocate and, in fact, is a legitimate corrollary of
our physical theory. When the products of bacterial
action are regarded as resulting from the dynamic
energy of bacteria, then the quantity of the products will
vary with the amount of energy which they exert, and as
the kind and degree of energy result from the molecular
structure of the bacterium, it is necessary that this structure is further inquired into that its bearing on attenuation
may be understood. Assuming then that pathogenic products are formed from albuminoids whose molecular structure
have been changed by the dynamic energy of pathogenic
bacteria. and that a bacterium is pathogenic only when its

molecular waves coincide in time and periods with those of
albuminoids of the body, and the latter are disrupted or
changed into pathogenic products by the former, it follows
that artificially induced changes in the molecular grouping
of a pathogenic bacterium, by which its molecular waves
are changed in time and periods, will weaken or destroy its
power of producing pathogenic products. The degree of
attenuation of a bacterium, i. e., the degree of its poison-
ous-product-producing-power will vary inversely with
the extent of that change in its molecular grouping by
which its molecular waves are made to vibrate in discord
with albuminoids that had previously vibrated in accord
with the bacterium. Attenuating agencies then, from our
point of view, are those which can change the molecular
structure of a bacterium, and thus weaken its power of producing its functional products, without injuring its power of
reproduction, growth or other physical or biologic qualities,
and, "that which distinguishes bacteria in their normal
state from those artificially debilitated" is this difference in
the molecular grouping of the protoplasmic contents of the
two micro-organisms. While "the extremely striking phenomena of attenuation" is explained as a dissonance of the
molecular waves of the bacterium and albuminoid, conson-
ance of these waves results in a disruption and conversion
of the albuminoid into pathogenic products, but when these
waves are dissonant this can not occur. Degree of attenuation would, of course, vary directly with the wave coincidence of the bacterium and albuminoid, and the duration
of attenuation which a bacterium acquires in this manner,
will depend on the permanence of the acquired change in
its molecular grouping; as a matter of fact it is known that
this varies with the nature of the micro-organism, the
means of attenuation, and the extent of its use

We have already called attention to the wondrous energy of etherial wave motions, and pointed out that they constitute au immense store-house of energy upon which nature relies to conduct her manifold operations; light, heat, electricity, etc., are simply wave lengths and nothing more. Sound is also a wave motion, and while it differs from ethereal waves in being propagated through the atmosphere only, yet its waves are known to be governed by laws which are strickingly similar to those which govern ether waves. As the laws of sound are better understood, and its waves more easily illustrated than the others, we have introduced a lengthy quotation from Prof. Helmholt'z essay on "Harmony in Music," which discusses the relation between waves of sound and music, and also serves to illustrate how either waves of the same length may differ in their energy, as those of sound do, from the height of their crests—and with similar results if sound can be regarded as manifested energy of sound-waves. It is established that height of wave crest corresponds to loudness of sound; if, then, the energy of waves produced by pathogenic bacteria differs in the same way, the crests of those produced by attenuated bacteria are not as high as those of bacteria not attenuated. Or these latter may differ from the former in time of periodic recurrence, and thus cause the difference in their physical properties; if the waves of attenuated bacteria are one-half wave length out of unison with those of virulent bacteria, then the waves of the former would have much less functional power than the others, just as the waves of sound given out by two forks whose vibrations are, in the same degree, out of unison. One will vibrate more rapidly than the other, and the sound produced by them will be less than would be produced by two forks vibrating in the same periodic time. In one case, attenuation of the bacteria changes its time of vibration and less pathogenic products—which represents its functional action—is produced. In the other case, the change of time of the vibrations of one of the forks destroys its coincidence with the other, and less sound—its functional product—is produced.*

When the three artificial methods named—by which immunity may be produced—are carefully analyzed, and the

---

*Extracts from a lecture "On the Physiological Causes of Harmony in Music," by Professor H. Helmoltz:

"First of all, what is a musical tone? Common experience teaches us that all sounding bodies are in a state of vibration. This vibration can be seen and felt, and in the case of loud sounds we feel the trembling of the air even without touching the sounding bodies. Physical science has ascertained that any series of impulses which produce a vibration of the air will, if repeated with sufficient rapidity, generate sound.

"This sound becomes a musical tone, when such rapid impulses recur with perfect regularity and in precisely equal times. Irregular agitation of the air generates only noise. The pitch of a musical tone depends on the amount of impulses which take place in a given time; the more there are in the same time the higher or sharper is the tone. And, as before remarked, there is found to be a close relationship between the well-known harmonious musical intervals and the number of the vibrations of the air. If twice as many vibrations are performed in the same time for one tone as for another, the first is the octave above the second. If the number of vibrations in the same time are as 2 to 3, the two tones form a fifth; if they are as 4 to 5, the two tones form a major third.

"If you observe that the number of the vibrations which generate the tones of the major chord C E G C are in the ratio of the numbers 4, 5, 6, 8, you can deduce from these all other relations of musical tones, by imagining a new major chord, having the same relations of the number of vibrations to be formed upon each of the above named tones. The numbers of vibrations within the limits of audible tones would be obtained by executing the calculation thus indicated, are extraordinarily different. Since the octave above any tone has twice

rationale of the dissimilar means employed is investigated from our point of observation, it will be seen that these processes, which appear to be quite different, are in fact practically one and the same; the poisonous albumins (tox-albumins) formed from albuminoids by the dynamic energy of pathogenic bacteria are the sole cause of immunity. When this condition is induced by introducing virulent bacteria into the animal body, a large amount of tox albu-

---

as many vibrations as•the tone itself, the second octave above will have four times, the third has eight times as many. Our modern pianofortes have seven octaves. Their highest tones, therefore, perform 128 vibrations in the time that their lowest tone makes one single vibration.

  *   *   *   *   *   *

"The musical pitch of a tone depends entirely on the number of vibrations of the air in a second, and not at all upon the mode in which they are produced. It is quite indifferent whether they are generated by the vibrating strings of a piano or violin, the vocal chords of the human larynx, the metal tongues of the harmonium, the reeds of the clarionet, aboe and bassoon, the trembling lips of the trumpeter, or the air cut by a sharp edge in organ pipes and flutes.

"A tone of the same number of vibrations has always the same pitch, by whichever one of these instruments it is produced. That which distinguishes the note A of a piano, for example, from the equally high A of the violin, flute, clarionet, or trumpet, is called the quality of the tone, and to this we shall have to recur presently.

"You will perceive, from what has been hitherto adduced, that the human ear is affected by vibrations of the air, within certain degrees of rapidity,—viz: from about 20 to about 32,000 in a second,—and that the sensation of musical tone arises from this affection.

"That the sensation thus excited is a sensation of musical tone, does not depend in any way upon the peculiar manner in which the air is agitated, but solely on the peculiar powers of sensation possessed by our ears and auditory nerves. I remarked, a little while ago, that when the tones are loud the agitation of the air is perceptible to the skin. In this way deaf mutes can perceive the motion of the air, which we call sound. But they do not hear, that is, they have no

mins are elaborated and the toxic effects of these are acute-
ly manifested in the production of the symptoms of the disease, but when these substances are added in small and frequently repeated doses, or by inoculations with attenuated
microbes, which is practically the same thing, immunity
will result as certainly as by the first method, but the toxic
effects of the immunizing agents are but slightly manifested;
in other words, the toxic symptoms will vary directly with

------

sensation of tone in the ear.   They feel the motion by the nerves of
the skin, producing that peculiar description of sensation called
whirring.   The limits of the rapidity of vibration within which the
ear feels an agitation of the air to be sound, depend also wholly upon
the peculiar construction of the ear.

*   *   *   *   *   *

"I must now describe the propagation of sound through the atmosphere.   The motion of a mass of air through which a tone passes, belongs to the so-called wave motions—a class of motions of great importance in physics.   Light, as well as sound, is one of these motions.

"The name is derived from the analogy of waves on the surface of
water, and these will best illustrate the peculiarity of this description
of motion.

"When a point in a surface of still water is agitated,—as by throw-
ing in a stone,—the motion thus caused is propagated in the form of
waves, which spread in rings over the surface of the water.   The circles of waves continue to increase even after rest has been restored at
the point first affected.   At the same time the waves become continually lower, the further they are removed from the center of motion,
and gradually disappear.   On each wave-ring we distinguish ridges
or crests, and hollows or troughs.

"Crest and trough together form a wave, and we measure its length
from one crest to the next.

"While the wave passes over the surface of the fluid, the particles
of the water which form it do not move on with it.   This is easily
seen, by floating a chip or straw on the water.   When the waves
reach the chip, they raise or depress it, but when they have passed
over it, the position of the chip is not perceptibly changed.

"Now, a light floating chip has no motion different from that of the

the quantity of toxic agent introduced; when this is large the symptoms produced are grave, when the dose is small, the symptoms are mild. This explanation of the philosophy of attenuation, it will be seen, is in line with that which we have made of fermentation, and infection, and, as will be shown later on, with that of immunity. All these processes are grounded primarily in molecular dynamics, and secondly, in the principles of chemistry and bio-

---

adjacent particles of water. Hence we conclude that these particles do not follow the wave, but, after some pitching up and down, remain in their original position. That which really advances as a wave is, consequently, not the particles of water themselves, but only a superficial form, which continues to be built up by fresh particles of water. The paths of the separate particles of water are more nearly vertical circles, iu which they revolve with a tolerably uniform velocity, as long as the waves pass over them.

"To return from waves of water to waves of sound. Imagine an elastic fluid like air to replace the water, and the waves of this replaced water to be compressed by an inflexible plate laid on their surface, the fluid being prevented from escapiug laterally from the pressure. Then on the waves being thus flattened out, the ridges where the fluid had been heaped up will produce much greater density than the hollows, from which the fluid had been removed to form the ridges. Hence the ridges are replaced by condensed strata of air, and the hollows by rarefied strata. Now, further imagine that these compressed waves are propagated by the same law as before, and that also the vertical circular orbits of the several particles of water are compressed into horizontal straight lines. Then the waves of sound will retain the peculiarity of having the particles of air only oscillating backwards and forwards in a straight line, while the wave itself remains merely a progressive form of motion, continually composed of fresh particles of air. The immediate result then would be waves of sound spreading out horizontally from their origin.

"But the expansion of waves of sound is not limited, like those of water, to a horizontal surface. They can spread out in any direction whatsoever. Suppose the circles generated by a stone thrown into

logy. This explanation of attenuation, so far as I know, is the only one which attempts to explain its philosophy; and when we consider the importance of the subject, and realize that until correct answers to Frankel's interrogatories are made there can be no scientific knowledge how attenuated bacteria confer immunity, it seems strange that this subject has not heretofore received that scientific treatment that its importance demands. We attach great im-

———

the water to extend in all directions of space, and you will have the spherical waves of air by which sound is propagated.

"Hence we can continue to illustrate the peculiarities of the motion of sound, by the well-known visible motions of waves of water.

"The length of a wave of water, measured from crest to crest, is extremely different. A falling drop, or a breath of air, gently curls the surface of the water. The waves in the wake of a steamboat toss the swimmer or skiff severely. But the waves of a stormy ocean can find room in their hollows for the keel of a ship of the line, and their ridges can scarcely be overlooked from the mast-head. The waves of sound present similar differences. The little curls of water with short lengths of wave correspond to high tones. Thus the contrabass C has a wave thirty-five feet long, its higher octave a wave of half the length, while the highest tones of a piano have wave of only three inches in length.

"You perceive that the pitch of the tone corresponds to the length of the wave. To this we should add that the height of the ridges, or, transferred to air, the degree of alternate condensation and rarefaction, corresponds to the loudness of the tone. But waves of the same height may have different forms. The crest of the ridge, for example, may be rounded off, or pointed. Corresponding varieties also occur in waves of sound of the same pitch and loudness. The so-called 'timbre' or quality of tone is what corresponds to the form of the waves of water. The conception of form is transferred from waves of water to waves of sound. Supposing waves of water of different forms to be pressed flat as before, the surface, having been leveled, will of course display no differences of form, but, in the interior of the mass of water we shall have different distributions of pressure, and hence of density, which exactly correspond with the differences of

portance to that part of this subject which relates to duration of attenuation, and believe that the philosophy of this
is nearly allied to that of duration of immunity; in both
cases this condition is variable. It is a much discussed
question whether acquired characteristics, or those artificially induced, are ever inheritable, but there can be no question that the physical and functional qualities which go
with attenuated bacteria are transmissible from parent to

---

form in the still uncompressed surface. In this sense, then, we can
continue to speak of the form of waves of sound, and can represent it
geometrically. We make the curve rise where the pressure, and
hence density, increases, and fall where it diminishes—just as if we
had a compressed fluid beneath the curve, which would expand to the
height of the curve in order to regain its natural density.

*   *   *   *   *   *

"Finally, I would direct your attention to an instructive spectacle,
which I have never been able to review without a certain degree of
physico-scientific delight, because it displays to the bodily eye, on the
surface of water, what otherwise could only be recognized by the
mind's eye of the mathematical thinker in a mass of air traversed in
all directions by waves of sound. I allude to the composition of
many different systems of waves, as they pass over one another, each
undisturbedly pursuing its own path. We can watch it from the par-
apet of any bridge spanning a river, but it is most complete and sublime when viewed from a cliff beside the sea. It is then rare not to
see innumerable systems of waves, of various lengths, propagated in
various directions. The longest come from the deep sea and dash
against the shore. Where the boiling breakers burst, shorter waves
arise, and run back again towards the sea. Perhaps a bird of prey'
darting after a fish, gives rise to a system of circular waves, which,
rocking over the undulating surface, are propagated with the same
regularity as on the mirror of an inland lake. And thus, from the
distant horizon, where white lines of foam on the steel-blue surface
betray the coming trains of waves, down to the sand beneath our feet,
where the impression of their arcs remains, there is unfolded before
our eyes a sublime image of immeasurable power and increasing variety, which, as the eye at once recognizes its pervading order and
law, enchains and exalts without confusing the mind.

progeny through many generations of these micro-organisms, and it will be shown, later on, that the changes produced in the albuminoids of the animal body by immunizing agencies, are likewise transmitted from parent to progeny, but in this case they are not continued beyond the life of the animal organism. If, now, we are correct in the assumption that the energy of a substance is derived from its molecular waves, then a change of energy, in degree

---

"Now, just in the same way you must conceive the air of a concert-hall or ballroom traversed in every direction, and not merely on the surface, by a variegated crowd of intersecting wave systems. From the mouths of the male singers proceed waves of six to twelve feet in length; from the lips of the songstresses dart shorter waves, from eighteen to thirty-six inches long. The rustling of silken skirts excites little curls in the air, each instrument in the orchestra emits its peculiar wave, and all these systems expand spherically from their respective centers, dart through each other, are reflected from the walls of the room, and thus rush backwards and forwards, until they succumb to the greater force of newly generated tones.

"Although this spectacle is veiled from the material, we have another bodily organ, the ear, specially adapted to reveal it to us. This analyses the interdigitation of the waves, which in such cases would be far more confused than the intersection of the water uudulations, separates the several tones which compose it, and distinguishes the voices of men and women,—nay, even of individuals,—the peculiar qualities of tone given out by each instrument, the rustling of the dresses, the footfalls of the walkers, and so on.

"It is necessary to examine the circumstances with greater minuteness. When a bird of prey dips into the sea, rings of waves arise, which are propagated as slowly and regularly upon the moving surface as upon a surface at rest. These rings are cut into the curved surface of the waves in precisely the same way as they would have been into the still surface of a lake. The form of the external surface of the water is determined in this, as in other more complicated cases, by taking the height of each point to be the height of all the ridges of the waves which coincide at this point at one time, after deducting the sum of all similarly simultaneous coincident hollows.

or kind, is the result of a change in the molecular grouping of the substance in question. If, for example, the substance is a virulent pathogenic bacterium, the energy or
ability to disrupt and convert certain body albuminoids into
tox-albumins, is derived from the molecular waves, and
when this bacterium becomes attenuated, its energy becomes
weakened, and its molecular waves do not have the same
degree of power of elaborating poisonous albumins because

Such a sum of positive magnitudes (the ridges) and negative magnitudes (the hollows), where the latter have to be subtracted instead of
being added, is called an algebraical sum. Using this term, then, we
may say that the height of every point of the surface of the water is
equal to the algebraical sum of all the portions of the waves which at
that moment there concur.

"It is the same with the waves of sound. They, too, are added together at every point of the mass of air, as well as in contact with the
listener's ear. For them also the degree of condensation and the velocity of the particles of air in the passages of the organ of hearing
are equal to the algebraical sums of the separate degrees of condensation and of the velocities of the waves of sound, considered apart.
This single motion of the air produced by the simultaneous action of
various sounding bodies, has now to be analyzed by the ear into the
separate parts which correspond to their separate effects. For doing
this the ear is much more unfavorably situated than the eye. The
latter surveys the whole undulating surface at a glance. But the ear
can, of course, only perceive the motion of the particles of air which
impinge upon it. And yet ths ear solves its problem with the greatest exactness, certainty, and determinacy. This power of the ear is
of supreme importance for hearing. Were it not present, it would be
impossible to distinguish different tones.

"Some recent anatomical discoveries appear to give a clue to the explanation of this important power of the ear.

"You will all have observed the phenomena of the sympathetic
production of tones in musical instruments. The string of a pianoforte, when the damper is raised, begins to vibrate as soon as its
proper tone is produced in its neighborhood with sufficient force by
some other means. When this foreign tone ceases, the tone of the

attenuation has changed the normal grouping of its molecules.   The change in molecular grouping of the bacterium,
by attenuating causes, is more or less permanent according
to the nature of the bacterium, the attenuating means used,
and the duration and extent of exposure of the bacterium
to the attenuating influence.

In the foregoing pages we have outlined the advanced
views of science regarding the nature of matter, motion

---

string will be heard to continue some little time longer. If we put
little paper riders on the strings, they will be jerked off when its tone
is thus produced in the neighborhood. This sympathetic action of
the string depends on the impact of the vibrating particles of air
against the string and its sounding-board. Each separate wave-crest
(or condensation) of air which passes by the string is, of course, too
weak to produce a sensible motion in it. But when a long series of
wave-crests (or condensations) strike the string in such a manner that
each succeeding one increases the slight tremor which resulted from
the action of its predecessors, the effect finally becomes sensible. It
is a process of exactly the same nature as the swinging of a heavy
bell. A powerful man can scarcely move it sensibly by a single impulse. A boy, by pulling the rope at regular intervals corresponding
to the time of its oscillations, can gradually bring it into violent motion. This peculiar reinforcement of vibration depends entirely on
the rhythmical application of the impulse. When the bell has been
once made to vibrate as a pendulum in a very small arc, and the boy
always pulls the rope as it falls, and at a time that his pull augments
the existing velocity of the bell, this velocity, increasing slightly at
each pull, will gradually become considerable. But if the boy apply
his power at irregular intervals, sometimes increasing and sometimes
diminishing the motion of the bell, he will produce no sensible effect.

"In the same way that a mere boy is thus enabled to swing a heavy
bell, the tremors of light and mobile air suffice to set in motion the
heavy and solid mass of steel contained in a tuning-fork, provided
that the tone which is excited in the air is exactly in unison with that
of the fork, because in this case also every impact of a wave of air
against the fork increases the motions excited by the like previous
flows.

and energy, and have presented reasons for our belief that, to the molecular structure of pathogenic bacteria is to be ascribed the dynamic energy whereby they are enabled to convert susceptible albuminoids into disease-producing albumins; and further, that the difference between suscep·tible and immune albuminoids consists in their susceptibil-ity to, or immunity from the wave impacts of pathogenic bacteria. Pathogenesis is then a result of two factors, viz.: that of the bacterium, its molecular structure and re-sulting wave motions, and that of the albuminoid molecules. of the body, their molecular structure and resulting wave

---

"This experiment is most conveniently performed on a fork, which is fastened to a sounding-board, the air being excited by a similar fork of precisely the same pitch. If one is struck, the other will be found after a few seconds to be sounding also. Then damp the first fork, by touching it for a moment with a finger, and the second will continue the tone. The second will then bring the first into vibra-tion, and so on.

"But if a very small piece of wax be attached to the ends of one of the forks, whereby its pitch will be rendered scarcely perceptibly lower than the other, the sympathetic vibration of the second fork ceases, because the times of oscillation are no longer the same in each. The blows which the waves of air excited by the first inflict upon the sounding-board of the second fork, are indeed for a time in the same direction as the motions of the second fork, and consequent-ly increase the latter, but after a very short time they cease to be so, and consequently destroy the slight motion which they had previous-ly excited.

"Lighter and more mobile elastic bodies, as for example strings, can be set in motion by a much smaller number of aerial impulses. Hence they can be set in sympathetic motion much more easily than tuning-forks, and by means of a musical tone which is far less accu-rately in unison with themselves.

"Now, then, if several tones are sounded in the neighborhood of a pianoforte, no string can be set in sympathetic vibration unless it is in unison with one of those tones. For example, depress the forte

motions; when the wave motion of a bacterium coincide
with those of an albuminoid molecule, a disruption of this
will result from the successive wave impacts of the bacterium, and the constituent molecules of the albuminoid will
be liberated from their chemical bonds and left free to recombine into other substances.   If, now, these new-formed
poisons produce the toxic symptoms of infection, then
this bacterium is pathogenic.   If, on the contrary, the
animal body does not contain albuminoid molecules which
vibrate in unison with the bacterium a disruption will not
occur, poisonous albumins will not be formed, and the bac-

---

pedal (thus raising the dampers), and put paper riders on all the
strings. They will, of course, leap off when their strings are put in
vibration. Then let several voices or instruments sound tones in the
neghborhood. All those riders, and only those, will leap off which
are placed upon strings that correspond to tones of the same pitch as
those sounds. You perceive that a pianoforte is also capable of analyzing the wave confusion of the air into its elementary constituents.

    *   *   *   *   *   *

"We have hitherto spoken only of compositions of waves of different lengths. We will now compound waves of the same length which
are moving in the same direction. The result will be entirely different, according as the elevations of one coincide with those of the
other (in which case elevations of double the height and depressions
of double the depth are produced), or the elevations of one fall on the
depressions of the other. If both waves have the same height, so
that the elevations of one exactly fit into the depressions of the other,
both elevations and depressions will vanish in the second case, and
the two waves will mutually destroy each other. Similarly two
waves of sound, as well as two waves of water, may mutually destroy
each other, when the condensations of one coincide with the rarefac-
tions of the other. This remarkable phenomenon, wherein sound is
silenced by precisely similar sound, is called the interference of
sounds.

    *   *   *   *   *   *

"Interference leads us to the so-called musical beats. If two tones

terium will not be pathogenic for this animal. But as the albuminoid molecules of the same animal differ in molecular structure, some of these will be disrupted by one class of bacteria, others by other bacteria, and other molecules of an almost indefinite series, by varieties of bacteria whose molecular structure is equally variable; the products which result from the action between a pathogenic bacterium and a susceptible albuminoid molecule will then vary with the microbe and the albuminoid concerned. A more marked difference of molecular structure is believed to exist between animals of different classes or species, consequently a bacterium of

---

of exactly the same pitch are produced simultaneously, and their elevations coincide at first, they will never cease to coincide, and if they did not coincide at first, they never will coincide. The two tones will either perpetually reinforce or perpetually destroy each other, but if the two tones have only approximatively equal pitches, and their elevations at first coincide, so that they mutually reinforce each other, the elevations of one will gradually outstrip the elevations of the other. Times will come when the elevations of the one fall upon the depressions of the other, and then other times when the more rapidly advancing elevations of the one will have again reached the elevations of the other. These alternations become sensible by that alternate increase and decrease of loudness, which we call a beat. These beats may often be heard when two instruments which are not exactly in unison play a note of the same name. When the two or three strings which are struck by the same hammer on a piano are out of tune, the beats may be distinctly heard. Very slow and regular beats often produce a fine effect in sostenuto passages, as in sacred part songs, by pealing through the lofty aisles like majestic waves, or by a gentle tremor giving the tone a character of enthusiasm and emotion. The greater the difference of pitches, the quicker the beats. As long as no more than four to six beats occur in a second, the ear readily distinguishes the alternate reinforcements of the tone. If the beats are more rapid, the tone grates on the ear, or, if it is high, becomes cutting. A grating tone is one interrupted by rapid breaks, like that of the letter R, which is produced by interrupting the tone of the voice by a tremor of the tongue or uvula.

well known pathogenic qualities is not virulent to all animals alike; to some it is actively so, to others—it may be
of the same family—it is less virulent or harmless. Again,
a bacterium that is virulent to a susceptible animal is not
virilent to this animal after it has been made immune by
any of the methods named.   These important facts cannot
be explained when we regard virulence of a bacterium as
a fixed quality of the microbe, e. g., as its secretion or excretion, but when virulence is regarded as a result of the
harmonious action of two factors—the microbe and albuminoid molecule—which are both equally necessary—these
phenomena are easily explained.

When we consider the nature of an albumin molecule;
its massiveness and molecular complexity; its instability
and lines of weak union, and the purpose it serves in the
animal economy, we must, I think, concede that the views
which have been given of pathogenesis are correct, provided the evidence which we have furnished is conclusive that
bacteria owe their special energy to their molecular struct·
ure.   A more detailed discussion of these subjects makes
this statement more evident.   For example: it is believed
that an albumin molecule is comprised of many simpler
molecules, and these of many atoms, and, that the nature
and number of the atoms, their method of grouping to form
simple molecules and the group arrangement of these, determine the chemical and physical nature of an albumin
molecule.   (2) While albumin in its different isomeric
forms contains probably the same number and kind of
atoms, and probably agrees, in the main, in the grouping
of molecules, it is believed that differences in this respect,
which may be slight, are the cause of those which are known
to exist between different varieties, isomeric forms of this substance. (3) An albumin molecule has lines of weak union, like

the lines of cleavage in crystals, along which it is most easily broken, often into poisonous substances. "At least three distinct series of chemical bodies are thus formed, viz: an acid series, an aromatic series, and a basic series. Out of the innumerable products arising from the action of bacteria cells upon albumin molecules, and which have been extracted and studied, will be mentioned iodol, cresol, and skatol in the aromatic series. Creatine in the basic, and uric acid in the acid series, serving only as mere examples." Aside from these, a large number of tox-albumins are known to be formed from broken or changed albuminoid molecules and (4) When we remember that albumin comprises the principal bulk of animal tissues, and that it is the pabulum from which the cells and tissues of the body receive their matter and energy; that to one it furnishes bone, to another muscle, and to others, nerve, brain, cartilage, blood and the various glandular elements, it seems evident that no substance having a uniform structure and form of energy could do this. On the contrary, the molecular structure and forms of energy must be as varied almost as that of protoplasm from which the tissues of the body are mainly constructed. Altogether, we are forced to believe that albuminoid molecules of the body differ in many ways, and in innumerable instances in molecular structure and dynamic energy. When now a similar diversity of structure is found to exist in baceria, it is plain to be seen what will result when these are brought into contact with albuminoid molecules. If the wave-motions of the bacteria coincide with those of any of the albuminoid molecules, a disruption of the weaker— the albuminoid—will occur, and a recombination of the liberated molecules will produce pathogenic products; such albuminoids are susceptible to the dynamic energy of the bacteria, and such bacteria are pathogenic. But if a bacterium does not find albuminoid molecules which they can disrupt, then the albuminoids are immune from this bacterium, and the bacterium is innocuous to the organism.

We have previously explained why pathogenic products inhibit the bacteria which produce them and finally arrest the disease when the products accumulate in sufficient quantity; the waves of the pathogenic products necessarily antagonize those of the bacterium. But their action does not end here, they are, not only inhibitory bodies but are active agents which, also, produce changes in the albuminoids. For illustration: the molecular waves of a bacterium must coincide with those of albuminoid molecules before the former can convert the latter into pathogenic products; if now the dynamic energy of these products can influence the molecular waves of the bacterium, they would equally influence other substances having waves of the same set of periods; this we have seen is the condition of susceptible albuminoids—their wave-motions must coincide with those of the bacterium before the former can convert the latter into pathogenic products. Such albuminoids being less stable than pathogenic products, will be changed in their molecular structure by the molecular wave impacts of these substances, and as change of molecular structure involves a change of molecular vibration, such albuminoids would cease to be susceptible to the bacterium in question, and the organism, say man, would be immune from that disease of which this bacterium is an etiological factor. The change in the molecular structure of the susceptible albuminoids, produced by the dynamic energy of the pathogenic products, is very similar to, if not identical with that produced by attenuation in the molecular structure of virulent bacteria; a regrouping of constituent molecules, and a corresponding change of specific energy, in degree or direction, occurs alike in both substances—the bacterium and albumin molecule.

The change in the molecular grouping of bacteria protoplasm produced by attenuating agencies, which we will call the condition of attenuation, is known to be variable in both degree and duration. If we regard the ability of pathogenic bacteria to produce pathogenic products as a functional power, then a weakening of this power by attenuating agencies, is equivalent to the same degree of weakness in a production of pathogenic products, and consequently, the ability of these microbes to produce pathogenic products will vary directly with their degree of attenuation; from virulent bacteria on one side, to bacteria having the faintest pathogenic power on the other, represents the extremes between which are found the various stages at which attenuation may be arrested. But the changes produced in bacteria by attenuating agencies, are not confined to the limits which restrict the pathogenic power of these microbes; the attenuation of bacteria may be carried beyond the limit required for the complete loss of pathogenic power, and this, too, without, in the least, changing the physical appearance of the bacteria, or their habits of growth or reproduction. Hence, we believe that attenuation consists of a change in the molecular grouping of the bacteria (to which they owe their pathogenic power) and, when this change is radical the change in dynamic energy will be equally radical. We have learned, furthermore, that the condition of attenuation is more or less permanent in its duration and, in fact, becomes a normal characteristic that is transmitted by inheritance from parent to progeny through many generations of the microbe. Some bacteria hold these changes of attenuation more tenaciously than others, and again, some attenuating agencies, and the length of time they are permitted to act, also determine the duration of attenuation.

If, now, the molecular changes produced in albuminoid molecules by the dynamic energy of pathogenic products, belong to the same order of phenomena as those produced in bacteria by attenuating agencies, we feel justified in our belief that such albuminoid changes will also vary in degree and duration, i. e., they will become normal characteristics of the albuminoid molecules, and will be transmitted from parent to progeny for variable degrees of time. Now, if immunity of the organism consists in immunity of its albuminoid molecules from pathogenic bacteria, and if this varies in degree and duration with the nature of the bacterium and albuminoid molecule concerned, it becomes evident why infectious diseases are so variable in the degree and duration of immunity which one attack confers from other attacks of the same disease.

Continuing this line of investigation, we must follow up the probable history and final disposition of the albuminoid molecules which have been artificially changed in molecular structure. The complexity of these bodies, their location in the liquids or solids of the body, the interdependence of the various tissues, racial, individual and environmental influences, and our imperfect knowledge of these subjects, make it impossible to follow the changes produced in albuminoid molecules through all their ramifications and intricacies, but it is sufficient for our purpose that the probable result of such changes in albuminoid molecules will depend whether they are harmful or harmless, and whether they occur in the fluids or solids of the body. If they are harmful to the organism it is probable they are not retained; this appears to be the disposition of pathogenic products, those which result from a disruption of susceptible albuminoids by pathogenic bacteria.   If they are harmless, and further, if such albuminoid molecules have not lost their nutritive qualities, they are probably retained and give immunity from the causative bacterium and the disease which it produces.   Or under certain conditions, for example, those in which albuminoid molecules have been disrupted by the pathogenic products, it may be that the new albuminoids formed by a recombination of the liberated molecules, will have qualities antagonistic to the pathologenic products, i. e., will have antitoxic properties. The same principles of molecular physics and chemistry by which pathogenic products inhibit, or antagonize pathogenic bacteria, give albuminoid molecules formed from other albuminoids that had been disrupted by pathogenic products, power of antagonizing these products. Now, it is a question of fact that the blood and body juices of animals made immune from certain infectious diseases actually contain such antitoxic substances; in fact, they have been extracted and isolated from the blood serum, and found to have valuable curative, as well as protective qualities.   Our knowledge of these antitoxic bodies, which are formed in the blood of animals made immune, is, however, of too recent date to pronounce on the frequency of their occurrence, or their therapeutic value.   Thus far they have been discovered in the blood of animals made immune from three diseases; they were first discovered in the blood of animals vaccinated against tetanus and diphtheria (Behring and Kitasato), and afterwards in the blood of animals made immune from pneumonia (G. and F. Klemperer).   And, it is claimed, the blood serum of animals immune from tetanus possesses eminent therapeutic properties; that when this is introduced subcutaneously into the bodies of animals which have the initial symptoms of tetanus, these will speedily disappear and the animal will be cured of this malady which, hitherto, has resisted treatment; three cures of this disease in man, by this method, have already been reported.   These brilliant discoveries have stimulated workers in bacteriology to renewed efforts, and already reports of extensive work in this direction have been made, but, so far, they have not added to the list of diseases which can be arrested by inoculations of immune blood. On the contrary, Metschnikoff has recently reported a series of experimental investigations in hog cholera, conducted in his usual painstaking and admirable manner, the result of which is that the blood and blood serum of animals suffering from this disease, or artificially made immune from it, does not contain antitoxic properties.   Why these substances are formed in the blood from some diseases and not from others, is   question, in the present state of our knowledge of these matters, that can not be positively answered.   But it is probable that this depends on whether the albumin molecules are disrupted and recombined into antitoxic bodies, or simply changed in molecular grouping without being disrupted. In the latter case, even though the change is slight, the molecules would have the periods of their vibrations changed, and consequently would no longer be influenced by the successive wave impacts of the bacterium, i. e., they would be converted into immunizing bodies.   Antitoxic bodies, it will be observed, have the property of antagonizing the bacterial poisons and also of resisting the wave impacts of the bacteria, while, on the other hand, immunizing bodies possess the latter quality only.

The difference in chemical and therapeutic properties of the products that are elaborated by the same bacterium, furnishes evidence that these differ in molecular structure and dynamic energy, and it is probable, therefore, that they affect albuminoid molecules quite differently; one product may cause a disruption of these molecules, and another simply a molecular rearrangement; one would then form antitoxics, the other immunizing bodies, or, it is possible, both may be formed in the same disease. But, when we consider the probability of the retention within the body of these substances, the chances are that the one which has received the least change, the immunizing substance, would be longest retained. The other would, probably, be quickly eliminated. Now, as a question of fact, this appears to be what really occurs, and it furnishes an explanation, how immunity of sucklings is induced by the mother's milk after she had been immunized by frequent inoculations with the bacteria products of tetanus bacilli. (Brieger and Ehrlich.)*

Albuminoid molecules of the solid or stable tissues of the body are much less vulnerable to pathogenic bacteria than are those of the fluids or unstable tissues. But once invaded, for illstration, by microbes of tuberculosis, leprosy, actinomycosis, or cancer, the malady produced is chronic; one attack gives no immunity, and such diseases are not self-limited in their duration. Syphilis and erysipelas, it is true, are not thus conditioned, but they are not strictly confined to the fixed or stable tissues of the body. These facts are those which the physical theory would lead us to expect; the highly specialized structure of the cells of solid tissues, and their fixed qualities, render them less vulnerable to the dynamic energy of bacteria, and, whatever change occurs in their molecular structure, from this influence, would destroy their functional use and make them

---

* Deutsche Medicinische Wochenschrift. Translated for "Texas Sanitarian" by Dr. J. O. Lewright.

harmful to the organism.   Again, the products resulting
from such change, which comprise the immunizing agents,
do not have sufficient energy to affect the albuminoids of
normal cells so highly specialized and, consequently, cannot change the periods of their vibration; their field of
action is confined to cells which have already been weakened by the pathogenic bacteria.   Let us, for example, examine the effect of tuberculin—the pathogenic product of
tubercle bacilli—on cells which have been weakened by the
bacilli.   It is now well known that when this substance is
subcutaneously introduced into the body of, say a tubercu-
lous individual, that those cells already weakened by tubercle
bacilli, will quickly undergo coagulative necrosis.   Such tissue, of course, is immune from further influence of the bacilli, and to this extent the pathogenic product gives immunity,
but this is strictly a local effect and does not extend beyond
the area of destroyed and harmful tissue.   If these products
could change the molecular structure of healthy cells without imparing their useful qualities, then it would give general immunity from the bacterium, and the disease would
be self-limited.

Immunity is not alone artificially induced by pathogenic
products; while these unquestionably are the agents which
most frequently bring about this condition—the reason for
which has already been given—there is evidence that it
may be artificially produced by other means which, although very unlike in chemical properties, are believed to
act through the dynamic energy of their molecular waves
changing the structure and wave motions of susceptible albuminoid molecules of the organism.   The following facts
furnish examples of this class.   "Hueppe and Wood found
that a species of bacteria, clearly distinct from anthrax bacillus, apparently innocuous and strictly saprophytic, was

able to secure even very susceptible animals, such as mice and guinea pigs, against anthrax;'' Roux and Chamberland secured immunity from different diseases by inoculating with one kind of bacterium, and Wooldridge has been able to produce this condition—immunity from anthrax— by inoculating with normal tissue juices*

This is the proper place to consider certain discoveries which are more or less connected with the subject of immunity. Behring made the rat, which is naturally refractory to anthrax, susceptible, by diminishing the alkales-

---

*''Immunity from Anthrax by Injection of Chemical Bodies,'' by Dr. L. C. Wooldridge: Dr. L. C. Wooldridge recently communicated to the Royal Society a method by which he had been able to protect rabbits from anthrax, which is of considerable interest in connection with the general question of the nature of protection in this and other diseases depending on micro-organisms. The method consists in cultivating the anthrax-bacillus in an alkaline solution of a peculiar proteid body, which can be obtained from the testis and thymus gland. The growth of the microbes is not abundant, and after two days, at 37° C., they are removed from the culture fluid by filtration. A small quantity of the filtered liquid is injected into the circulation of a rabbit and the animal can then withstand the inoculation of extremely virulent anthrax blood. The bacillus, itself, grown in this culture fluid, has no protective influence. It either kills or it has no effect. The result is extremely curious. For, hitherto protection against zymotic diseases has been effected by the communication to the animal of a modified form of the disease against which protection is sought. In Dr. Wooldridge's experiments the protection must be produced by some chemical body, the product of the activity of the bacillus. The observation belongs to the new order of facts and appears to fall in with Pasteur's theory as to the method in which immunity to hydrophobia is produced by inoculation of the spinal cord of rabbits. Both find some support in Prof. Cosh's experiments with perchloride of mercury, in which it was shown that after animals had taken a sufficient quantity direct, they were no longer liable to anthrax.''—Brit. Med. Journal, 5, 11, 1887.

cence of its body. Emmerich, Pawlowsky, Bouchard, and Freudenreich repressed a commencing anthrax infection and cured it by introducing other micro-organisms, such as the erysipelas coccus, the micrococcus prodigiosus, and the bacillus pyocyaneus.*

---

*"The History of Microbean Products which favor Infection," by Prof. Charles Bouchard, Paris, France. (1) It is not uncommon to see two species of microbe invade at the same time an animal organism; the result of this mixed infection may be widely dissimilar. Sometimes the two pathogenic agents develop side by side, without any reciprocal influence upon each other; sometimes the animal will find in one of them an unexpected ally against the other, when it would have succumbed to one affection, it survives the combined attack of the two; finally, the two microbes, acting together, sometimes overcomes an organism which would have successfully resisted either one of the invaders, if attacking alone. The writer remarks that this is the result most frequently seen; almost always infection intensifies infection.

It is not necessary that the two microbes both be pathogenic in order that their results become virulent; one of them may be a simple saprophyte. It sometimes happens, in fact, that two forms, neither of which are pathogenic to the animal, produce, when associated, death. This has been demonstrated by experiments upon what has been called symptomatic anthrax. The product of this microbe, so virulent to the bovine species, is without action upon the rabbit; inoculation of this animal, either under the skin or under the muscles, produces with this bacillus no result.

Let us now take another microbe, inoffensive of itself—the bacillus prodigiosus—and mingling it with the microbe of charbon symptomatique, inject it into the tissues or under the skin, the animal dies and we find at the point of injection a tumor with all the characteristics of anthrax. We here find two microbes, neither of which are pathogenic for the rabbit when injected alone, but which, when taken together, give to one of them a quality, which, in relation to this animal, it had not before. The action of the auxiliary microbe, in this case the bacillus prodigiosus, depends upon the matters secreted. The result is the same whether we employ the living cultures, the

Immunity, prodnced by these unusual means, and the unexpected results obtained by introducing different bacteria into the organism at the same time, have given rise to considerable speculation as to the rationale of their action, and have tested the ingenuity of theorists to harmonize them with preconceived hypotheses. But when immunity is conceived to be that molecular structure of the albuminoids of the body which renders them invulnerable to the wave impacts of a given bacterium, it becomes apparent that a substance having the requisite molecular structure

---

sterilized cultures, or simply the glycerole of the cultures, that is, what has been called the extract.

The power of one microbe to modify the qualities of another, does not belong, especially, to the bacillus prodigiosus; other forms possess analogous properties. This fact has been established by experiment, by injecting the soluable matters of the staphylococcus aureus, of the proteus vulgaris, or of the sterilized extracts of putrefying flesh. The facts which have just been stated seem to have a general application, as may be demonstrated by experiments with a large number of forms. For instance, Fluegge and Vissokovitch have shown that certain soluble products render possible the development in the organism of non-pathogenic microbes. Grawitz and Barry have found the inoculation of the staphylococcus aureus with the secretions of the bacillus prodigiosus intensifies the pus-producing power of the former. Recently Monti has rendered virulent the attenuated cultures of the pneumococcus and streptococcus by injecting at the same time the secretions of other microbean forms, and especially of the proteus vulgaris.

The soluble matters which favor infection do not act by altering locally the tissues into which they are introduced; their action is general upon the whole organism. In fact, the better methods where we wish to diminish or abolish the resistance of an animal to invasion, is to inject the soluble products directly into the blood. In this way we obtain the effect with doses twenty or thirty times less, and much more rapidly than when the injection is made at the point of inoculation.—Gazette Hebdomadaire, July, 1890. Journal Am. Med. Association, Aug. 30, 1890.

to convert susceptible into immune albuminoids, would, for the same reason, have immunizing power; and, while the pathogenic products of bacteria most frequently possess this structure—because of the influences under which they were formed—still, other substances, e. g., Wooldridge's tissue febrinogen, may also have it and, in this case would be an immunizing substance. Again, different kinds of bacteria may be able to change the structure of susceptible albuminoid molecules by attacking them at some of their lines of weak union; perhaps one will find its point of attack at one line, and others at different points, the result would be the same, the molecule would be changed in its molecular structure and, consequently, would be immune from the molecular bombardment of the bacterium. Or, immune albuminoids, by similar agencies, may become susceptible, and the immunity of animals may thus be destroyed.

Before ending our presentation of the phenomena of acquired immunity, and the various means by which this condition can be induced, we desire to discuss a subject of prime importance which is intimately connected with this condition, and concerning which we entertain views that are opposed by those usually held. The subject is that which relates to the manner in which pathogenic products are formed. We have already given reasons for our belief that these products result from a recombination of the molecules of albuminoids which have been disrupted by the dynamic energy of pathogenic bacteria. This is a cardinal principle of the physical theory, but is not the view that is generally accepted. It is held, almost universally, by others, that pathogenic products are secretions or excretions of pathogenic bacteria. Frankel voices the opinion generally entertained, in the following quotation from his "Bacteriology:" "We may conveniently summarize our opinions by saying: The *pathogenic bacteria are those which, by their vital action, produce excretions injurious to the bodies of men and animals.*"

This hypothesis is an outgrowth of the enzyme theory of fermentation which, it will be remembered, assumes that ferment bacteria secrete enzymes or unorganized ferments, which are the active causes of fermentation. We do not think it necessary to again point out the weak points, or, more correctly, the fallacies of this hypothesis, but in view of its bearing on the causes of pathogenesis, we will again invite your attention to the experimental work of Helmholtz, Mitscherlich and Hoffman, which has been previously referred to, and rely upon the evidence, which they furnish, as conclusive of the incorrectness of the enzyme theory. If this theory of fermentation is not supported, but, on the contrary, is disproved by the evidence, then it is more than probable that it has nothing to do with pathogenic products. We believe this hypothesis is unscientific and contrary to sound analogy. It is unscientific in ascribing functional qualities to structureless and undifferentiated protoplasm of bacteria cells that are possessed only by complex and highly specialized cells of multicellular organisms. It is contrary to the analogy furnished by that strikingly similar process in fermentation, wherein ferment products are formed from a recombination of the molecules of the fermentable substances which had been liberated from their chemical bonds by the dynamic energy of ferment bacteria; the ferment products, collectively, consist of the same elements and in the same proportion which compose the fermentable substance. Analogy would, therefore, teach that pathogenic products are similarly formed from a recombination of albuminoid molecules which had been liberated from their chemical bonds by the dynamic energy of pathogenic bacteria.

But aside from these reasons for not accepting this hypothesis, others will arise when we view the phenomena of pathogenesis from this standpoint; for example: the behavior of leucocytes of an immune, and of a susceptible organism towards pathogenic bacteria. It is well known that certain amœboid cells of the animal organisms,—of which the white blood-cells comprise the principal representatives,—in the performance of their physiological work, will attack and destroy dead and inert substances which find their way into the body; this scavenging characteristic has given these cells the name of phagocytes (scavengers). Now, it has been ascertained that innocuous bacteria, when introduced into the animal organism, are disposed of by its phagocytes notwithstanding such bacteria are virulent to other animals, and further, the phagocytes of animals which have been made immune from an infectious bacterium, say anthrax bacillus, will attack and devour this microbe. But the case is quite different with the phagocytes of susceptible animals; if the bacterium which is devoured by the phagocytes of an immune animal is introduced into the blood of the same kind of animal before it becomes immune, i. e., when it is susceptible, the phagocytes will no longer attack it, but on the contrary will hasten out of its way, and the animal becomes infected. In the recent discussion of "Phagocytes and Immunity," at the London Pathological Society, Dr. Wm. Hunter refers to this matter in the following language:

"Before discussing the probability of one or the other doctrine (phagocytic and humoral theories of immunity), it is well to have the facts clearly before our minds. The facts, then, to take the simplest case, are briefly these:

That in nonprotected animals virulent bacteria inoculated subcutaneously produce little or no local change, grow unhindered, and cause general infection, while in protected animals inoculation of the same bacteria is followed by a local leucocytosis at the point of inoculation, a certain degree of inflammation is set up, and general infection is prevented." It is to this fact that Frankel refers in his inquiry, "Why can bacteria in their natural condition grow and multiply in the bodies of susceptible animals, and not in the bodies of non-susceptible or immune animals?" Or when we accept the statement, almost universally admitted, that it is the products of pathogenic bacteria that give virulent qualities, we are confronted by the inquiry: Why are pathogenic products poisonous to susceptible and not poisonous to immune animals? It may be observed, in passing, that the difficulty of answering this question is not lessened by our knowledge of the fact that immune animals are not, by any means, immune from the pathogenic products in question. When the difference of behavior of phagocytes is viewed from the standpoint of those who believe that pathogenic products are secretions or excretions of pathogenic bacteria, difficulties arise which have not been overcome. But when this subject is viewed from the standpoint of the physical theory, it is seen that the products and not the bacteria influence the behavior of the phagocytes; the presence of these in the susceptible animal, paralyze the phagocytes and thus prevent them from attacking and destroying the bacteria, consequently they grow unhindered, and infection takes place. The absence of these poisons in the immune animal, allows the phagocytes the full possession of their scavenging power and the bacteria are destroyed, consequently they do not grow unhindered in the bodies of such animals, and infection does not take place. All this becomes plain when we regard pathogenic products as substances which have formed from the liberated molecules of susceptible albuminoids, i. e., albuminoid molecules which are susceptible to the dynamic energy of a given bacterium. These bodies, for reasons given, are found only in unprotected animals, and it is in these alone that the bacterium is virulent; the same bacterium is not virulent, but harmless to protected animals for the reason that they do not contain susceptible albuminoids, i. e., albuminoid molecules which the bacterium can disrupt and convert into pathogenic products.

CHAPTER VI.

THE EXHAUSTION THEORY OF PASTEUR AND KLEBS; THE
RETENTION THEORY OF CHAUVEAU, AND THE HUMORAL
THEORY OF BUCHNER AND BEHRING.

Webster defines theory as a "philosophical explanation
of phenomena." If we fellow this definition and require
that theories of immunity give a philosophical explanation
of their phenomena, it will best enable us to determine
which of these has the most value. But it is not sufficient
that the phenomena of immunity alone receive philosophical explanation; those of infection belong to the same
order, and immunity from infection cannot be explained
until the condition which produces susceptibility to infection is clearly understood. In other words, the laws of infection and those of immunity, imply conditions of the
organism which bear opposite relations to infectious bacteria. Or, if we represent the infectious bacteria and the
resistance of the organism as two forces, the bacteria would
be an attacking, and the tissue resistance of the organism
a defending force; one is complemental to the other, and
the nature of neither can be understood without an understanding of both. In our examination of the theories of
immunity we will, therefore, ascertain their competency to
explain the phenomena of infection and those of immunity.

The phenomena of infection are those of infectious diseases and comprise (a) the period of incubation which pre-
cedes an acute infectious fever; (b) the self-limited duration
of this class of diseases; and (c) the varying degree of ma-

lignancy of these diseases as manifested in different epidemics.

The phenomena of immunity comprise (a) natural immunity; (b) acquired immunity, (1) by means of single attacks, (2) by means of infectious bacteria, (3) by means of attenuated bacteria, (4) by means of the products of virulent bacteria, (5) by means of blood serum of immune animals, (6) by means of varying processes; (c) inoculation fever; (d) the nature and causes of those changes produced in virulent bacteria by attenuation, and (e) the different behavior of the leucocytes of (1) immune animals, and (2) susceptible animals toward virulent bacteria.

And finally, a philosophical explanation should be made, why the agencies which have been mentioned will secure immunity from some acute infectious diseases and fail to do this for other diseases of the same class! And why acquired immunity is not of equal duration for all infectious diseases!

The theories which we shall pass in review and submit to the examination test above noticed are:

1. The exhaustion theory of Pasteur and Klebs.
2. The retention theory of Chauveau.
3. The phagocytic theory of Metschnikoff.
4. The humoral theory of Buchner and Behring, and last, but we hope, not least,
5. The physical theory, for which we simply ask a patient hearing, and hope its obscure birth will not prejudice its claims. Give it a "free fight and an open field," and if it can not maintain itself in this gladiatorial contest against its illustrious competitors, backed as they are by the prestige of world renowned professional reputations, honorably and laboriously won, and what in the opinion of many Americans is more valuable, by the *eclat* of European

authority, then let it pass away, and be remembered only as an honest but misguided ambition, whose aim is to arrive at truth, and place medicine on a more scientific basis than that it now occupies.

THE EXHAUTION THEORY.

This theory is an old acquaintance which we had occasion to criticise in our discussion of the theories of fermentation. It is assumed by the advocates of this theory that when infectious bacteria are introduced within the animal organism, they obtain the oxygen they require for respiratory purposes from the organic molecules of its fluids, and, that the molecules which have been robbed of their oxygen fall to pieces, and their elements recombine into other substances,—the toxics of disease.

This simple and, at first glance, plausible theory of infection completely breaks down when it is required to explain known facts and phenomena of this process ; for example, some varieties of extremely virulent bacteria are strictly aerobic in habit, i. e., they obtain the oxygen they require from the air, and can not take it from organic molecules, consequently, they can not live when placed within the animal organism. The bacilli of diphtheria and tetanus are of this order, they are strictly confined to the surfaces of the body, and when inoculated they pass but a short distance from the place of introduction, yet they are extremely infectious and quickly produce toxic effects. Pasteur's theory cannot explain how these bacteria cause infection. As they can not live within the body they do not rob the organic molecules of its fluids of oxygen, and thus cause them to fall to pieces and recombine into infectious products.

But aside from this objection, which is fatal to Pasteur's

theory of infection, this hypothesis can give no satisfactory explanation why bacteria affect the organism differently. It would seem that pathogenic bacteria in their greed for oxygen would obtain this from molecules which would most easily yield it. Consequently, the products formed from a recombination of the molecules should display but slight variability, and as these are disease-producing substances, there should be but slight difference in the diseases which they produce. Bacteria, under these circumstances, would have no distinctive qualities, and their action in the organism would be confined to that of abstracting oxygen from organic molecules; hence, this theory can not explain the difference in symptoms and pathological lesions of infectious diseases, except that which would result from the qualities of the organic molecules from which the bacteria obtain oxygen, and as these would be limited to those which most easily yield this element, the types of the resulting diseases would be equally limited in number.

This theory is equally vulnerable when attacked from its immunity side. For illustration, it teaches that immunity from a pathogenic bacterium is a result of certain changes of a nutritive character in the composition of the blood. Or, to be more explicit, it is claimed by the advocates of this theory that, when a given pathogenic bacterium is introduced into the blood of a susceptible animal, it feeds upon certain nutritive elements of the blood, upon which it continues to grow and multiply, until the pabulum is finally destroyed—eaten up, when the bacterium perishes from starvation and the disease of which it is the cause is then arrested. The animal is thereafter immune from this bacterium for the simple reason that it can not find nourishment within the body on which it can live. The duration of the disease will then depend on the length of time required for the bacteria to exhaust the pabulum; while the duration of immunity is dependent on a reformation of the destroyed pabulum. If this is never reformed immunity is permanent.

It is urged against this theory that chemical analyses of the blood of immune and susceptible animals fail to show difference in chemical composition. Hence the changes in the composition of the blood, which this theory assumes must occur, particularly in an animal which has been invaded by different kinds of pathogenic bacteria, are found not to exist. The theory, therefore, is not warranted by facts.

Its explanation, that the absence of pathological phenomena during the incubative stage, and their mildness during the inoculative stage are due to the paucity of bacteria introduced, cannot be regarded as entirely satisfactory, much less so the explanation, that the varying malignancy of cases, whether of the epidemic or sporadic form, is caused by the vital energy of the bacteria. This cannot be accepted until the causes which determine this vital energy are philosophically explained.

A serious if not a fatal objection is its absolute failure to explain how immunity is induced by inoculations with the products of bacterial action. In this case no bacteria are introduced into the body, consequently, none of the nutritive elements of the blood are destroyed, yet immunity can be as certainly secured by this method as by that of using infectious bacteria for protective inoculations. It fails also to explain the mild type of symptoms which results from inoculations with attenuated bacteria, when contrasted with the severe symptoms produced by virulent bacteria. Experimental evidence has conclusively proven that attenuation does not lessen the vigor of growth of bacteria, nor does it weaken their power of appropriating nutritive materials which they require for their growth and multiplication, therefore it cannot be claimed that the mildness of symptoms which follow inoculations with attenuated bacteria is caused by feebleness of their appropriative or growing qualities. The theory is unable to explain how blood-serum of immune animals gives immunity, and by what method one variety of bacteria gives immunity from several diseases, or how this condition may be brought about by means of substances which are in no way related to pathogenic bacteria.

It cannot explain natural immunity unless it assumes without warrant, that certain nutritive substances which are normal in the blood of susceptible animals, and presumably necessary, are not present in the blood of the same animals after they have been made refractory. And finally, it has no explanation to offer of the difference in the behavior of leucocytes of (1) immune animals, and (2) refractory animals toward virulent bacteria.

THE THEORY OF RETENTION.

This hypothesis is based upon the assumption that the products of bacterial action are retained more or less permanently in the body of the previously infected animal. It is conceded, I believe, by all theories, that the products of bacteria and not the bacteria themselves are the cause of infectious diseases, and it is generally recognized that these substances in some way inhibit the pathogenic activity of the causative bacteria, and are intimately concerned in the production of immunity.

This is the view taken by the advocates of the retention theory. They claim the products of a bacterium, when formed in sufficient amounts, will inhibit the further pathogenic activity of this microbe; that in fact, the bacterium is not only inhibited by its products, but in some cases is destroyed by the same means. Therefore, if it be established that these products remain permanently in the bodies of the previously infected, then many of the most important phenomena of acquired immunity would be simply and clearly explained.   Many persons are not, however, satisfied with this simple explanation and cannot accept its conclusions; e. g., the fact that acquired immunity extends for years, and it may be for life, requires that the ptomaines of such diseases remain in the body during this long period of time. Or, if the individual has passed successfully through several of the acute infectious diseases, it requires that the ptomaines of each of these diseases remain as harmless agents in the body of the individual.   As the products of bacteria are chemical substances of an alkaloid or albuminoid nature, we are at a loss to understand how they can remain so persistently in the body when other, similar substances, are quickly eliminated from it.   If the chemical products of bacteria can be retained in the body for an indefinite time, then an individual, who has successfully passed through attacks of several different kinds of infectious fevers, would carry about in his blood the chemical poisons of each of these diseases and, notwithstanding the fact that these poisons are deadly to others, they are assumed to be absolutely harmless to the individual.   If this theory is true, and immunity from an infection is caused by retention of poisons in the blood of a previously infected individual, it follows as an inevitable result that the individual would be immune, and in equal degree, from the bacterium and its poison; but this is proven to be not the case; immunity from a bacterium does not give equal immunity from its poisonous products.

But the premises upon which this theory is constructed

are, in part, unquestionably true. For example: it is established by irrefutable experimental evidence that the products of bacterial action, and not the bacteria themselves, are
the cause of the phenomena and pathological lesions of disease, and that these products, in certain proportional
amounts, will inhibit further action of the bacteria, i. e.,
will arrest the process by which these products are formed.
Therefore, if it can be shown that these products are retained in the body of a previously infected individual, then
some of the phenomena of acquired immunity are easily
and simply explained; but unfortunately this can not be
done. This vital point, which is so necessary to the integ-
rity of this hypothesis is not supported by evidence; on the
contrary, it is at variance with well established physiological facts; those, for illustration, which occur when poisons
are taken into the body. It is well known that these are
not retained for any considerable length of time, and the
tolerance acquired by the habitues of opium, alcohol, etc.,
to these drugs, is not due to their retention within the
body, but to profound changes produced by them in its
tissues. These illustrations do not, therefore, furnish that
support to the retention theory which is claimed for them.
The fact that immunity from a virulent bacterium does not
necessarily give immunity from its poisonous products,
when considered in connection with that taught us by the
introduction of large doses of tox-albumins, which, instead
of immunizing, kill the animal, surely indicates that these
substances when introduced, in at first small and then in
gradually increasing doses, give immunity to the animal—
not by their direct action and, therefore, retention, but
likewise by producing profound changes in the tissue elements of the body; this condition—which the physical

theory explains—and not the tox-albumins, is the immediate causes of immunity.*

THE HUMORAL THEORY.

"In all our researches on immunity we are confronted with the question, How is it that immunity is conferred for one disease only, or certainly for one series only? Why should an attack of vaccinia or variola be necessary to insure protection against another attack of variola? The exhaustion of assimilation theory was early found insufficient to account for it, as it was also to explain the different actions or the bacillus prodigiosus, which, whilst diminishing the action of a dose of anthrax, increases the virulence of Chauveau's bacillus, or charbon symptomatique. The local immunity theory in the same way was found insufficient,

---

*If an animal be inoculated with the secretions of certain microbes, as for instance the bacillus pyocyaneus or of symptomatic anthrax, and if several days afterwards the same animal be inoculated with the living microbe, the previous inoculation having been sterilized, there will be no infection, the animal survives, it has acquired an immunity; in other words, it has been vaccinated.

If allowing several days to escape between the time of the introduction of the secretions of the microbe and the living bacterium, the two are introduced at the same time, the animal succumbs, in fact, death is produced more surely and more rapidly than if the inoculation is made with the living microbe alone. It thus appears that the bacterian products have two diametrically opposite effects, depending upon their time of introduction as related to the inoculation of the pathogenic form.

This conclusion clearly follows from the experiments conducted by Bouchard with the bacillus pyocyaneus. He says, I had thought that immunity would be obtained more quickly by injecting en block the vaccinating matters at the beginning of the malady, than by awaiting for this immunity to result from the gradual development of the pathogenic agent. I had imagined that in this manner one would be

and local inflammation as a factor in immunity, has had its day, perhaps, however, to be revived at no distant date. The adaptation theory or cellular chemical theory advocated by Klebs and Grawitz, Roberts' theory of massing leucocytes, and Metschnikoff's phagocytic theory have in turn ousted the retention theory advocated by Chauveau and Wernich, whilst the serum bactericidal theory has come forward to rout this if possible. It may be that we may have to cull something from each of several of these various hypotheses, but we must always bear in mind that the potential forms and functions of animal protoplasm are as great as the number of directions in which it actually develops.

------

able to diminish the duration of the disease, and possibly find in these products of bacteria a remedy for the infection itself. Experience has not justified this conjecture. The chemical matters produced by the bacillus pyocyaneus, which cut short or prevent the disease when injected a few days ago or a few weeks before the inoculation of the bacillus itself, if injected at the same time as the living form, that is, along with it or nearly at the same time, instead of diminishing the intensity of the infection, increases it and hastens the death of the animal.

The result is similar with symptomatic anthrax. The matters secreted by the microbes of this disease confer immunity against the inoculation practiced several days afterwards. If introduced into the system at the same time as the microbe, they favor the infection, and even render this bacterium pathogenic to animals ordinarily refractory, as for instance, the rabbit. Considerable quantities of the soluble matter secreted by this microbe injected into the veins of the rabbit produce no harm; inoculation of the animal with the living microbe is followed by no morbid symptoms. If injection into the veins and of the living microbes into the muscles be practiced at the same time, the animal dies in from twenty to forty-eight hours with an enormous tumor, in fact with anthrax.—Gazette Hebdomadaire, July. 1890, Journal Am. Med. Association, Aug. 30, 1890.

"In marked opposition to Metschnikoff's phagocytic theory, it was early brought forward by Emmerich and Mattei that the property of destroying bacteria rested in the fluid constituents of the blood. They first maintained that these *antibiotic* fluid constituents were the result of the activity of the bacteria themselves; in fact, they took up the retention theory. After further observation, however, they receded from this position, and came to the conclusion that the *antibiotic* substances were formed by the living cellular elements. Following them, Flügge, Nutall, and Nissen, who were afterwards joined by Buchner and others, described a bacteria killing substance which was present in the blood of immune animals, not only while it was circulating in the vessels, but also after it had been withdrawn and had practically undergone coagulation, the serum being separated from the corpuscles—in which condition the blood might be said to have lost all traces of vitality. They hold, therefore, that a certain quantity of toxic material, which apparently has little or nothing to do with the blood as a living tissue or fluid of the body (in this differing from Von Fòdòr's original idea, which was that there existed in the blood plasma a vital element which killed off the microorganism.) Following this up, Buchner maintains that the cause of recovery from an attack of an infective disease is directly due to a bacteria-killing action of the serum, and he further holds that phagocytosis is brought about merely by the dead microbes giving up their proteins, which acting chemically in a positive manner, draw the leucocytes to the point at which the organisms are breaking down; here the leucocytes take them up and ultimately digest and get rid of them, so that he looks upon phagocytosis merely as the result, first, of chemotaxis, and, secondly, of an effort on the part of the organism to get rid of devitalized microbes.*

Natural or acquired immunity of the body, it is claimed by the humoral theory, which has been fairly outlined in the above quotations, results alone from certain antibiotic, (bacteria killing)—or antitoxic (ptomaine destroying) substances contained in its blood serum or tissue juices; these may be there as a result of inheritance,—natural immunity —or artificially produced,—acquired immunity. It is further assumed by some of this school that an additional immunizing quality of the blood serum, that of attenuating virulent microbes which it cannot destroy, is a means and in some cases, the only protection the individual has from infectious bacteria.

The origin, or the how and why these various substances are produced in the blood serum is an unknown problem, about which there has been much speculation and but little demonstration. Its failure to explain this important matter marks one of the weakest points in the humoral theory. Speculation regarding this subject has, however, not been wanting, the outcome of which is, that the immunizing substances of the blood serum and other fluids of the body, are believed to be derived, either from the vital action of the bacteria (ptomaines, toxines or tox-albumins), the protoplasm (bacteri-protein), of dead bacteria, or from a reaction between bacteri-protein, the bacterial products, and the cellular elements of the organism.

Having now presented the principal features of the humoral theory, we will examine the evidence upon which this theory is based and, at the same time, will present the coun-

---

*G. Sims Woodhead, M. D., Discussion on Phagocytosis and Immunity at London Pathological Society.—Brit. Med. Jour.—Med. and Surgical Reporter.

ter-evidence which has been offered ,by opposing theories. In this way, it is believed, a clearer and more correct opinion, respecting its merits, may be arrived at, and the facts, thus sifted out and their relation to the causes of immunity, can be better interpreted.

Starting with the fact to which Fòdòr first called attention, but which is at present admitted on all sides, that a destruction of bacteria does take place in the bodies of living animals, we will examine how this is explained by the humoral theory.

" Petruschky discovered that the blood of animals insusceptible to anthrax, such as frogs, killed anthrax bacilli even when all cellular portions were carefully excluded. Behring showed that the serum of white rats, even when separated from the body, possessed this capability. Nutall was able to prove the same with regard to the aqueous humor, the ascites fluid, and other juices of the body, and this bacteria-destroying power of the blood serum has obtained a more universal importance from the independent, though simultaneous, labors of H. Buchner and Nissen. The results of their investigations may be summarized in a few words, as follows: Germ-free serum kept for several days at a low temperature has the power of killing bacteria-germs in a very short time. It is true that this capacity has its limits. If one inoculates more than a certain quantity of micro-organisms some will be killed, but for the survivors, the serum, instead of remaining a hostile element, becomes a source of aliment and the bacteria begin to increase in it.

" The different species show considerable differences of behavior in this respect, while some are peculiarly sensitive and sure to perish if brought into contact with serum, others are not in the least affected by it; and between these two groups is another, in which at first a slight check to development is observed, but which the bacteria soon get over and then proceed to grow and increase. The germ-killing, disinfecting power lies exclusively in its plasma, the cellular parts of the blood, the red and white corpuscles even, counteract and paralyze it. Under the influence of high temperatures, it quickly disappears, as already noticed; it is also diminished by the blood being left standing for a length of time, but repeated freezings and thawings do not affect it.

"Both investigators see a connection between this peculiar property of the serum and the processes which contribute principally to the coagulation of the blood. This, however, does not tell us much about the exact nature of the active principle, but a subsequent series of investigations by Buchner found that the germ-killing power of serum depends on the salt which it contains; a diminution of its saltness is accompanied with a diminution of its power to kill bacteria. One might, for a moment, be tempted to suppose that it is the mineral ingredients which, directly and immediately, serve as a basis of disinfecting power in the serum. Yet Buchner shows clearly that such cannot be the case, and that the salts are only of importance because they stand in intimate relation to the albuminoid matter of the blood, *the quantity and quality of which is decisive, and turns the scale one way or the other. The salts serve as solvents, or agglutinants, to the albuminates, and an altogether special, peculiar, and as yet unexplained condition of the serum albuminoids, is the active cause of the bacteria-killing agency.*"*

Many persons do not accept the conclusions which others

---

*Frankel's Text-Book of Bacteriology.   Italics mine.

believe are established by this experimental work. It is urged that blood serum which has, for some time, been removed from the body, and which, consequently, has lost its vitality and undergone other changes of a chemical nature, may be, and in some cases it is proven to be quite different in its behavior to bacteria from that of living normal blood; that serum thus conditioned does not fairly represent that in the vessels of the body, and results obtained with this cannot, in like manner and in the same degree, be obtained by using normal serum. In support of this view, it is stated that there is no constant relation between the degree of protection the body has from infectious bacteria, and the antibiotic or germ-killing power of its serum; that frequently the serum of animals which are quite sensitive to infective bacteria, after removal from the vessels will develop marked antibiotic powers. Thus it has been shown by Roux and Metschnikoff that the serum of young rats sensitive to anthrax has a bactericidal action on anthrax bacilli outside the body; and by Lubarsch that the extra vascular blood of a rabbit kills millions of anthrax bacilli, while a few of these injected into the blood suffice to kill the animal. Behring and Nissen admit that as regards the bacteria of pneumonia, anthrax and diphtheria, the blood serum and body juices of the same animals manifest the same degree of bactericidal power regardless of whether such animals are susceptible to, or have been made immune from these diseases.

Further evidence that living blood serum or body juices do not, as a rule, have bactericidal properties, is furnished by Metschnikoff and Ruffer. The former placed anthrax bacilli, wrapped in tissue paper or animal membrane, under the skin of frogs, which are naturally immune to anthrax; the latter placed the bacteria of charbon symptomatique, similarly prepared, under the skin of an immunized rabbit; it was found in both cases that the bacteria thus placed, notwithstanding they were constantly subject to the influence of the body juices, continued to grow and multiply, and after remaining in the tissues for forty-eight hours these bacteria had not lost, in the least degree, their infective qualities.

It is also claimed that certain infectious bacteria become attenuated when grown in the blood of refractory animals. "The bacillus anthracis when grown in the blood of vaccinated sheep no longer kill rabbits; the bacteria of erysipelas, when grown in the blood of vaccinated rabbits, lose their virulence; and the serum of immunized animals attenuates the bacillus pyocyaneus." As these results, like the former, were obtained with blood and serum which had for some time been removed from the body, they are open to the same objections which are made against the claim that normal and extra-vascular blood, alike possess antibiotic properties. But this matter is not left to speculation, it is proven that normal blood of immune animals does not always have attenuating properties. "Take, for instance, a rabbit vaccinated against anthrax and inoculate it with anthrax bacilli, thus allowing these to exist directly within the refractory organism. Such bacilli as are not destroyed preserve their virulence for a sufficiently long period, and it is possible to kill a guinea pig with a drop of exudation taken from the region of infection thirty hours after subcutaneous inoculation, eight days after inoculation into the anterior chamber of the eye. A sojourn of so long duration within the vaccinated organism, then, had not deprived the microbes of their virulence, although twenty-four hours suffices to completely attenuate the bacilli cultivated in the removed blood of vaccinated sheep. Years ago it was established in M. Pasteur's laboratory that the refractory organism, instead of being a favorable soil for the preservation of virulence, tends rather to reinforce this property. To exalt the virulence of an attenuated micro-organism, one always employs, not animals very susceptible to the specific disease, but those which are slightly susceptible, or it may be, under many circumstances, refractory. In this manner the most active anthrax virus has usually been obtained by passage through birds, notably fowls; the greatest virulence of chicken cholera was gained by passage through the vaccinated cock, and quite recently M. Malm has shown that passage of the anthrax bacillus through the organisms of dogs, which of all animals are the most refractory in this respect, increases the virulence in a most remarkable manner."*

The remarkable power the living blood of refractory animals has of increasing the virulence of bacteria, admits of another, and a more probable explanation, than that implied by Metschnikoff in the lecture, an extract from which we have quoted. Any considerable number of bacteria will contain individuals having different degrees of resisting, growing, and pathogenic powers. When, now, these, say pathogenic bacteria, are made to grow in an unfavorable food medium, for example, such as the body of a refractory animal is presumed to furnish, the weaker bacteria, those which have the least staying qualities, would perish, while those only in which these inherent qualities are highly developed, would survive to become the breeders of future bacteria possessing exalted powers of growth and functional

*Lecture on Phagocytosis and Immunity, by Dr. Elias Metschnikoff (Pasteur's Institute, France), British Medical Journal; Bacteriological World.

action.   When we consider that the life of a bacterium is
very brief, and that generation upon generation of these
micro-organisms will come and go in the short period of
hours, it will be seen that there might result from the above
conditions, a strain of bacteria having increased powers of
virulence.   At the same time this process of natural selection does not exclude the possibility that other pathogenic
bacteria may be differently affected by such unfavorable
conditions of life, and that the survivors of such bacteria
may themselves become attenuated by these conditions,
when compared with the same bacteria differently and more
favorably placed.

The latest and most important evidence in favor of the
humoral theory which has yet been advanced, is that furnished by recent discoveries of anti-toxic substances in the
bodies of immune animals, in certain cases.   Unlike antibiotic substances of immune animals, which are thought to
destroy pathogenic bacteria, the antitoxics are believed to
neutralize and make harmless the poisonous products of
bacteria without necessarily destroying the bacteria themselves.   Dr. Wooldridge, in 1888, described a method of
what he termed "chemical protection" against anthrax.
He found that when he administered tissue febrinogen, prepared from the thymus gland or from testicles, to rabbits,
they became immune from anthrax.   Hankin, taking up
this line of investigation, found that the body of the white
rat which is refractory to anthrax, contains a natural protective albuminoid which kills the bacillus of anthrax.   He
succeeded in isolating this substance, and proved that the
minute dose of it injected into mice suffering from anthrax
was sufficient to cure them of this disease.   Behring and
Kitasato, in 1890, obtained similar results in diphtheria and
tetanus from protective substances of like nature.   Kitasato

summarises his conclusions regarding tetanus as follows : "First, the blood of rabbits that have been immunized against tetanus possess properties which destroy tetanus poison; second, these properties have also been shown in extra-vascular blood, and in the serum of such blood; third, these properties are of such enduring nature that they remain effective in the organism of other animals, so that we are able to achieve eminent therapeutic action by the transfusion of this blood or serum; fourth, the tetanus poison-destroying properties are absent in the blood of animals that are not immunized against tetanus, and, if the tetanus poison is injected into animals not immunized, the same may be shown to exist in the blood and other fluids after the death of the animal."

Prof. M. Ogata, in connection with Mr. Iasuhara has succeeded in extracting from the blood of dogs, which are refractory to anthrax, a substance which he regards as a ferment; with this he has been able to immunize susceptible animals against anthrax. Messrs. Emmerich and O. Mastbaum claim to have protected susceptible animals and cured those infected with swine fever, by inoculations with tissue juices of immune animals. Messrs. Klemperer and Klemperer have arrived at the following conclusions from their experimental investigations upon pneumonia of animals :

"(1) Immunity against pneumonia can be bestowed upon susceptible animals by introducing in the tissues the sterilized products of growth of pneumococci. This immunity is, in general, but of temporary nature.

"(2) Such immunity induced by injection of bacterial products does not immediately manifest itself; indeed, fourteen days must elapse before the simple products bring about their effects. But if the sterilized products be heated either to 106° or 107.5° F. for three or four days, or to 140°

F. for two hours, then injections induce immunity within four days. The warmed "vaccine" leads to a reaction of but short duration; the unwarmed brings about a long continued febrile state, at the end of which the animal becomes immune.

"(3) The blood serum of a 'protected' animal injected into the veins of a susceptible animal confers immediate immunity.

"(4) What is more, this same substance has curative properties, acting, not so much on the pneumococci themselves (for in its presence these continue to proliferate), but upon the poisons or toxines manufactured by them."

It must be admitted that these remarkable results of experimental work furnish strong evidence that the blood and body juices of immune animals, in some cases at least, contain substances, which may be extracted and isolated, that have the power, not only of conferring immunity, but when properly used will sometimes arrest the infectious process and cure the disease. Even if we exclude as inconclusive that portion of this evidence obtained by using extra-vascular fluids of the body (on the assumption that results obtained with these are identical with those obtained with the living fluids of the body) there still remains sufficient evidence, which has not so far been disproved, that immunity is in some way connected with those substances which Hankin terms "defensive proteids," Buchner calls "alexins," Ogata and Iasuhara believe are "ferments," and all concede are derivatives of albumin.

It is true that substances having antitoxic and antibiotic power have, so far, been found in connection with but few infectious diseases, if they are formed in others of this class they have succeeded in eluding the diligent search which has many times been made for them. Until, therefore, it can be proven that "defensive proteids" are always produced in the bodies of previously infected animals, and, that these protective substances are inherent in the bodies of those naturally immune we are not warranted in believing that immunity is wholly due to anti-biotic or anti-toxic substances. Even then the humoral theory would lack that knowledge regarding how these substances are formed, and how they induce immunity, that is absolutely requisite to a scientific or philosophic explanation of phenomena.

It is not claimed that answers, such as they are, have not been made to these questions, but it is claimed that all such answers are either unsupported speculations, stated in those indefinite terms which convey no information as to the mode of action involved, or they are such that, if accepted, would destroy the humoral theory as a distinctive hypothesis, and place it with the retention theory of Chauveau.

The importance of a correct and clear understanding of these subjects becomes at once apparent when it is seen that such knowledge is necessary to a correct interpretation of those facts which relate to "defensive proteids" in the blood of immune animals, and their causative connection with immunity. The facts as stated must be recognized, however much we differ regarding their interpretation. If established that "defensive proteids" exist in the bodies of all immune animals,—inherently formed in the naturally immune, and acquired in the artificially immune,—as they are proven to be in anthrax, swine fever, tetanus and pneumonia of animals, the question of immunity, viewed from a practical standpoint, would be finally settled. But unfortunately the question is not thus settled, and it is a matter of doubt in many minds whether these substances in the diseases named should not be regarded as epiphenomena, or exceptions to the general rule, rather than phenomena necessarily and causatively related to immunity.

A critical examination of the different views regarding the manner of their formation, and that of their action will, I believe, enable us to better interpret their origin and significance. Beginning our examination then with the hypothesis that these substances are formed by a chemical reaction between the bacteri-protein of the dead microbes and certain cellular elements of the organism, we find certain serious, if not fatal, objections to this view. For illustration, this assumption can not explain how the "defensive proteid" against anthrax, which Hankin obtained from the naturally immune white rat was formed in its body. As the rat had never been invaded by the anthrax bacteria it will not do to ascribe its immunity to a chemical reaction between dead bacilli anthracis and its body cells. If we are to be guided in this examination by the admitted fact that nature is always uniform in her methods, we would certainly expect that the underlying principles, which she makes use of to form and function immunizing substances, would obtain alike in all cases, but it is evident that this rule will not apply in the hypothesis under examination; two widely separated methods are required, to account for these substances in the bodies of animals naturally, and those artificially immune.

Ehrlich has proved that susceptible animals can be made refractory to poisons which he extracted from the jequirity and castor oil bean; that obtained from the jequirity bean he calls *abrin*, and that from the castor oil bean he calls *ricin*; both of these substances are active poisons. It is said that one gramme of the latter is sufficient to kill one and a half millions of guinea pigs, but when these animals are given very small amounts of either of these substances with their food, in a short time they become immune from these poisons, and can then safely take doses subcutaneously or otherwise, the hundredth part of which would prove fatal to susceptible animals. In this example it is seen that bacteri-protein plays no part whatever in the acquired immunity; if *antiabrin* and *antiricin* are formed in the bodies of animals which have acquired immunity from the poisons *abrin* and *ricin*, as believed by Ehrlich, it is by some other method than a chemical or other reaction between dead bacteria (which are not present) and the tissue cells of the animal.

In this connection should be considered certain facts obtained by experimental work that belong to the same order of phenomena. Hueppe and Wood performed a successful protective inoculation—against anthrax — by means of a perfectly harmless species of bacteria, scarcely allied to anthrax bacillus. Hankin obtained from anthrax cultures a substance which granted immunity—an albuminoid body of peculiar qualities. Wooldridge took aqueous extracts, rich in albumen, from the thymus gland, and the parenchyma of the testicle of healthy animals, and with this obtained protection from anthrax. Emmerich, Bouchard, Pawlosky and Freudenreich repressed a commencing anthrax infection, and cured it by introducing other microorganisms, such as the erysipelas coccus, the micrococcus prodigiosus, and the bacillus pyocyaneous. It would be interesting to know how these facts can be harmonized with the belief that defensive proteids are derived from a reaction between bacteri-protein and the cellular elements of the organism.

The difficulties which we have noted are not overcome, nor is a clearer insight into the origin and mode of action of these substances obtained, by regarding them as a result

of chemical reaction between the products of bacteria and
the cellular elements of the organism.   The nature of the
supposed reaction, and of the substances involved, are simply
matters of speculation, there is really nothing known of any
of the supposed factors unless it be that relating to bacterial
products. But, even, if all the factors were known, and it
could be established by actual demonstration that immunizing proteids are formed in this manner, there would still
remain the same difficulty in explaining the cause of immunity that is found when we ascribe this condition to the
direct action of bacterial products.   In both cases alike, it
is necessary that the immunizing substances are permanently retained to explain permanency of immunity.   This does
not, however, accord with results obtained by Ehrlich in
his recent experimental investigations with *abrin* and *ricin*.
He has proven that when a mother mouse acquires immunity from either one of these poisonous substances that her
sucklings will also acqure immunity through the mother's
milk; in like manner immunity from tetanus may be induced in suckling mice through the milk of the immune
mother.   Now the explanation which is given of these
phenomena is, that the mother's milk contains defensive
proteids which, taken into the bodies of young mice, gave
them immunity. If this explanation is correct, it would follow that defensive proteids are not retained, but are eliminated from the body, and proves fatal to the hypothesis, at
least in its general application, that permanency of immunity is caused by permanent retention of defensive proteids.
   If we admit a causative relation between the products of
a bacterium as the efficient cause of a specific infection,
and the defensive proteid as the efficient cause of specific
immunity, and, further, admit that this defensive proteid is produced in the organism by a reaction between

the products of the bacterium and cellular elements of the
blood or body juices, we are confronted with the fact that
immunity against an infectious disease may be conferred by
substances that are not formed in this way; for example, by
tissue febrinogen from healthy animals, (Wooldridge),
and by bacteria in no way related to the infection, but
which, in fact, are absolutely non-pathogenic, (Hueppe and
Wood). In addition to these difficulties there is that of explaining the results obtained by Roux and Chamberland,
"who by inoculating with one kind of bacteria produced
immunity from several diseases, and who observed the frequent occurrence of reciprocal inoculative protection by
which one micro-organism secured immunity against another, and vice versa."*   If we seek to avoid these difficulties by denying that a causative and specific relation exists
between the products of a bacterium and the defensive proteid which confers immunity against the bacterium in question, it will then become necessary to explain why one form
of defensive proteid does not give immunity against all
pathogenic bacteria, or why immunity from small-pox does
not give protection against measles, whooping-cough or
scarlet fever.

We will next consider the methods by which these substances are supposed to confer immunity. These are, antibiotic or antitoxic; that is, the immunizing proteid arrests
and sometimes cures a beginning infection by either destroying the causative bacteria, or by neutralizing the effects
of its poisonous products. The first assumption must be
ruled out as it cannot explain how or why immunity is acquired from the products of bacteria. If these are really the
infectious agencies the disease could not develop until

---

*Frankel's Bacteriology.

these products are formed, and after this time the destruction of the bacteria by antibiotic substances will not explain why the disease becomes arrested while these disease-causing substances remain in the body, or, on the other hand, if they are eliminated from the body, why the animal continues to have immunity. It,therefore,became necessary to claim that the defensive proteids are not only antibiotic but also antitoxic substances, a claim which is strongly supported by the fact that the blood and serum of animals immune from tetanus can actually destroy both the tetanus bacilli and their poisonous products. But, unfortunately, this fact does not hold good in many other cases, and it is found that "in three diseases remarkable for their pronounced toxic character—vibrionic septicemia, pyocyanic disease, and hog cholera affecting the rabbit—as shown by the experiments of Charrin, Gamaleia and Selander, the toxines are so little attacked by the refractory organism that the same quantity of these substances (freed from bacteria) suffice to kill an animal very susceptible to one or other disease, and an animal vaccinated against it, and thus completely immune. So, too, non-fatal doses of these toxines produce in animals of the two categories the same febrile and inflammatory reactions. The proof is clear that there is no special destruction of toxines in the refractory animal, and that the 'toxicide property,' if it exists, is not one whit more developed after vaccination than before."*

We believe the difficulties in the way of an acceptance of the humoral theory, which have been named, arise from a false conception of the nature and origin of the "defensive proteids" of the body, and the relation these substances bear

---

*Lecture on "Phagocytosis and Immunity," by E. Metschnikoff.—British Medical Journal and Bacteriological World.

to immunity. For example, when the products of bacterial action are regarded as secretions or excretions of bacteria, and "defensive proteids" are regarded as chemical substances which have resulted from a chemical union of bacterial products, or of the substance of dead bacteria—bacteri·protein—with tissue elements of the body, the resulting products of this chemical union, the "defensive proteids," or what are termed antibiotic and antitoxic substances must be looked upon as substances foreign to the organism and the retention of such bodies for any considerable length of time is contrary to all experience, and involves a belief that is unphysiological. Again, if antitoxics destroy the poisons which produce disease by a chemical union with these substances, then this action would necessarily destroy the antitoxics as well; consequently we are required to believe that chemical substances, devoid of organic life, are capable of reproducing themselves. This, it is needless to say, is contrary to all biological experience. Finally, antitoxic substances have been found in the blood of animals immune from a few only of the many infectious diseases. We are not warranted, therefore, in regarding these bodies as the true cause of immunity until they are proved to exist in the blood, or body juices of animals made immune from at least a majority of these diseases.

When these subjects are examined in the light furnished by the physical theory, the difficulties which beset the humoral and other theories of immunity will disappear, and it will be explained how "defensive proteids" (antitoxics, etc.), may occur as *side products* of *immunization*, and, while they are immunizing and, it may be, antitoxic bodies, they are not the true causes of immunity. A vital distinction between the physical and other theories of immunity consists in the views entertained regarding how the products of bacterial action are formed. The physical theory claims that this process requires two factors, the bacterium and the albuminoids. Other theories claim but one factor, viz: the bacterium, which, it is taught, secretes or excretes its pathogenic products. The two factors required by the physical theory must bear a certain relationship towards each other, similar to that borne by ferment bacteria towards fermentable substances in order that the bacterium, by its dynamic energy can convert albuminoids into other (poisonous) forms of albumins—tox-albumins. A bacterium, then, is pathogenic only when its molecular waves vibrate in unison with the waves of certain albuminoids of the blood or body juices of the animal, and the former convert the latter into poisonous albumins.

The products of fermentation, those which result from a recombination of the molecules of the fermentable substance which were liberated from their chemical bonds by the dynamic energy—the wave impact—of ferment bacteria, have the power of antagonizing the energy of the bacterium, and it is well known that an accumulation of these in the fermenting solution, in a certain proportional amount, will arrest the fermentation, and this cannot be re-established until the excess of these products is removed from the solution. Now, it is believed that pathogenic products exercise the same influence over the energy of pathogenic bacteria that ferment products do over the energy of ferment bacteria; that is, that pathogenic products antagonize the energy of the pathogenic bacteria which cause them. The opposing influence of bacterial products to their causative bacteria, as we have already explained, is an opposing wave motion, and our reasons for believing that waves of bacterial products antagonize those of the bacteria, have also been given; but while the formation of ferment products is the final act of fermentation, such is not the case in pathogenesis; the role of pathogenic products is not confined to their inhibitory action. We believe that pathogenic products are themselves the cause of change in albuminoid molecules of the body of two kinds, viz: disruptive and reconstructive. When the molecular waves of a pathogenic product and those of albuminoids occur in the same periods of time, the · latter, which is less stable, will be disrupted by the former, and those substances which result from a recombination of the liberated molecules will have wave motions that antagonize or neutralize the former. (The principles of molecular physics which determine this result have already been stated.) In other words, these substances are antitoxic bodies; their energy opposes that of the toxalbumin and, in consequence of this quality, they are therapeutic agents which may cure a commencing disease of which the bacterium in question is the etiological cause. We cannot, however, believe that such substances have the power of self-multiplication; they are chemical bodies which are harmful to the organism, and analogy teaches that such substances are not long retained.

But a disruption of albuminoid molecules by pathogenic products, which requires a coincidence in time and periods of the wave motions of the two substances, is not the most frequent or important work performed by these products. That which results from their inhibitory or opposing influence, we believe, furnish the true cause of artificial immunity. The albuminoid molecules which are affected by this influence are those which are susceptible to the wave impacts of the causative bacteria. Both must vibrate in the same periods before the waves of the bacterium can disrupt the albuminoid molecules, consequently the resulting pathogenic products which antagonize the bacteria will, for the same reason, antagonize the set of albuminoid molecules which are vulnerable to the bacterial waves. But the influence of inhibitory waves is not so intense as that of disrupting waves; in one case they oppose, and in the other they supplement those of the albuminoids, consequently the inhibitory waves lose part of their intensity in overcoming the opposition of the albuminoid waves, and not until this influence is neutralized can they act directly on the molecular grouping in the albuminoid molecules. The dynamic energy of the pathogenic products which is thus left over, is not sufficient to disrupt the albuminoid molecules, but only enough to change the grouping of its molecular constituents, i. e., to give them another—isomeric—form of albumin. This change does not destroy the nutritive value of the albuminoid substance, nor its power of self reproduction, but it does change its molecular vibration and, in this way, renders it immune from that bacterium to which it had before been susceptible; that is, the animal is thereafter immune from the bacterium and the disease of which the bacterium is the cause, so long as the albuminoid molecules do not revert to their former structure.

If attenuating agencies change the molecular structure of virulent bacteria and thereby weaken, or, it may be, destroy their functional action without changing their mass appearance, their habits of growth and reproduction; that if, in fact, the molecular changes which attenuating causes impose on bacteria become a fixed habit of life which is transmitted by them to their progeny for varying periods of time, then it is surely not improbable that albuminoid molecules of the organism, which are much less stable in molecular structure, may be similarly changed by dynamic influences without changing their mass appearance or nutritive qualities, and this changed structure may become a fixed habit of the albuminoid molecule, as it is of the bacterium, which can be transmitted from parent to progeny for varying periods of time. Duration of immunity would then correspond to that group arrangement imposed on the albuminoid molecules by the dynamic energy of bacterial products. In both examples this condition is variable in duration; from a brief period, to permanency, represents the extreme limits between which are found those conditions occupied by different bacteria on one side, and different infectious diseases on the other.

The difference between susceptibility and immunity of animals to a given bacterium, will then depend upon the molecular structure of the albuminoids; those of the susceptible animal can be shaken apart and converted into ptomaines by the bacterium, because both vibrate in the same period of time, hence such albuminoids are susceptible. A different state of affairs exists in the organism of the immune animal; its albuminoids are not susceptible, do not vibrate in the same periods with the bacterium, and, consequently, are not disrupted and no ptomaines are formed. In this case the albuminoids are immune from the molecular bombardment of the bacterium to which in the other case the albuminoids are susceptible. It is thus seen that the question of immunity is resolved by this theory into one of molecular structure of the albuminoids. Those of immune animals have one molecular structure, those of susceptible animals have another, but in neither case are albuminoids changed in their chemical composition but, simply, in the grouping of their molecules; these form isomeric bodies of the same substance which, however, exhibit marked difference in their physical properties.

If now a pathogenic bacterium is pathogenic to a susceptible animal only, and not to an immune animal, i. e., does not produce ptomaines in such, it will be seen that the one thing needful to confer immunity from a bacterium, is that the albuminoids of the animal do not move in the same periods with those of the bacterium, and it is, therefore, not necessary that immune animals contain antitoxic or antibiotic bodies to protect the organism from infection.   In fact, if our conception of pathogenesis is correct, a pathogenic bacterium does not produce its poisons in the bodies of immune animals—those which are believed to contain antitoxic and antibiotic substances,—so that in the absence of toxines there is no necessity for antitoxines; nor does immunity in such cases require that antibiotics shall exist in the bodies of refractory animals for, in the absence of ptomaines, the scavengers of the body, its phagocytes are amply able to dispose of all bacteria against which the organism is immune.

CHAPTER VII.

THEORIES OF IMMUNITY, CONTINUED—THE PHAGOCYTIC THEORY—THE PHYSICAL THEORY.

THE PHAGOCYTIC THEORY.

The most popular theory of immunity, and that best supported by actual demonstration, of any which has yet been advanced, is the one which regards the white cell elements of the body as its defenders. These amœboid cells which, taken collectively, are termed phagocytes—scavengers—are regarded as trained warriors of the body which comprise its army of defense. "Metchnikoff, the chief expounder of this theory, claims that the destruction of the life of the bacteria by being taken into the bodies of certain cells,—phagocytes,—is the exclusive means which the organism makes use of in resisting the incursions of pathogenic bacteria." "An animal whose leucocytes can successfully battle with, and eat up a given species of bacteria, enjoys immunity from its deleterious effects."

If the animal body is regarded as an empire; the organs of the body as provinces; the blood-vessels and lymphatics as lines of commerce; its cells as individuals who are variously skilled in different industries, and the brain as the center of government, then, in accordance with this figure, we must regard the phagocytes as the standing army of the empire, divided into three separate *corps d' armee* or lines of defense.

Dropping now our simile and coming back to the animal

body, we find these lines of defense are, *first*, the skin and mucous membranes; *second*, the superficial lymphatic glands, and *third*, the deep lymphatic glands of the body. The leucocytes of the first line are those which make the first attack upon invading bacteria. When unsuccessful at the first line they fall back on the second and rally around the superficial lymphatics which become the next battle ground. If the leucocytes are still unsuccessful in the war they wage against the invading microbes, they again retire and form their last line of battle upon the deep lymphatics, where the final issue is made and it is there determined whether the microbes or the leucocytes are to be victorious; if the former, general infection of the body results; if the latter, the microbes are destroyed and the animal is saved.

It is thus seen that the Metschnikovian theory ascribes acquired immunity to "grim visaged war," a veritable battle of cells in which pathogenic bacteria are represented as an army of invasion, and the leucocytes as an army of defense. When the animal body is the battle ground of these warring cells, it is claimed that the fight is continued until it results in the death of the microbes if the leucocytes are victorious, or in the death of the animal if the microbes are the victors.

Two strange things are to be observed, in this war of cells, which have no parallel, at least in modern warfare. The first is found in the supposed means which the microbes employ in making an attack. They do not use Greek fire, mitrailleuse or gatling guns, but rely solely upon their own excreta with which to "paralyze" their foes.*

---

*The pathogenic bacteria are those which, by their vital action, produce excretions injurious to the bodies of men and animals.— FRANKEL'S BACTERIOLOGY.

The second consists in the behavior of the "trained warriors;" it is claimed that the leucocytes not only destroy pathogenic microbes but, cannibal like, actually devour them.

A theory of immunity which represents the body as a battle ground, and leucocytes and microbes as warring bodies whose war cry is "victory or death," may not be lacking in poetic fancy, or dramatic effect, but we submit that this conception is contrary to the simple methods which nature makes use of in carrying out her operations, and, therefore, is better suited to embelish the pages of fiction than it is to philosophically explain scientific phenomena.

We do not deny that phagocytes are scavengers of the body and that, in the performance of this physiological labor, they attack and destroy dead and inert substances. Nor do we deny that leucocytes of an immune animal will attack and devour bacteria which are known to be pathogenic to susceptible animals; this fact has been established by Metschnikoff by actual demonstration, and by experimental evidence of the most trustworthy nature, and it furnishes the principal support to the phagocytic theory of immunity. It is not, then, the fact, but its interpretation by the advocates of the phagocytic theory, to which we object. Our further contention is, that this theory extends physiological leucocytosis beyond its normal limits; it endows leucocytes with qualities closely allied to, if not identical with, conscious action and memory, and it claims that these complex mental qualities can be acquired by leucocytes, and can be transmitted by them to their progeny through indefinite generations of these cells. It is these features of the Metschnikovian theory that we contend are unscientific, and unsupported by evidence or sound analogy.

In submitting this theory to a critical examination we

must, first, determine what part of it is physiological and undisputed, and then separate this from the theory of immunity proper, which is purely speculative. The following extracts from Dr. G. Sims Woodhead's address on "Phagocytosis and Immunity" at the London Pathological Society, gives us a survey of the history and scope of what is known of physiological leucocytosis:

"Taking first the question of phagocytosis from its physiological standpoint, and quite apart from its bearing on immunity, it has long been recognized that inert particles finding their way into the animal system, or even introduced into the blood immediately after it has been removed from the body of the newt, say, are rapidly taken up by the leucocytes, and, in the body, from them, by the connective or other fixed tissue-cells. A similar process has been observed by every pathologist who has examined a healing wound, a section of the lung or the cells of the normal spleen. The significance of this process was first pointed out by Haeckel, who, comparing the leucocytes of the newt with those of the lower amœboid organisms, indicated that this must be the remaining evidence of a normal process of nutrition common to amœboid cells in whatever position they may exist; these cells, taking the vermillion or carbon particles, merely affording a better demonstration of what is going on in most amœboid cells.   *   *   *
All will accept it as proved that in the metamorphoses of the tadpole and of certain larval forms of insects, into fully developed forms, phagocytosis plays a most important part.
      *   *   *   *   *

"By a very natural extension, the part that this process of phagocytosis plays in the removal of a dead or effete tissue after injury or disease has been also traced, the destruction of the myeline sheath of nerve fibers by amœboid cells, the absorption of bone by the large multinucleated cells that lie in Howship's foveolæ or lacunæ, the similar cells observed in myeloid sarcoma, especially in various tumors of the jaw, where absorption of the bone is going on rapidly, the inception of fragments of yellow elastic fiber and the absorption of muscle, not only in the tadpole's tail, as observed by Metschnikoff and others, but also in the inflammatory process set up by the trichini spiralis, as worked out so beautifully by Soudakewitch and in the involuting uterus by Helme and others, have all been followed out."*

In 1881, Carl Rosser called attention to a power which the phagocytes of both animals and plants have of incepting, digesting and assimilating living micro-organisms, and he maintained as a result of his observations, that leucocytes of an immune animal can destroy bacteria which are themselves destructive to susceptible animals. Metschnikoff, working on the same lines, confirmed the earlier observations of Rosser, and from observing what appeared to be a contest between the white blood cells of the water flea and certain infectious micro-organisms (blastomyces), he was led to believe "that in every case of infection it is the white cell elements of the blood that, as scavenger-cells (phagocytes) have to save the organism if they can. If bacteria attack any part of the body, these cells, favored by their mobility at once appear at the place of danger and rush upon the invaders. If they are able to make the latter innocuous no infection takes place; if, however, these defenders of the organism struggle ineffectually and yield, the enemies begin to multiply and spread themselves over the unprotected domain." This ingenious and brilliant theory of immunity failed to meet with universal acceptance.

---

*Medical and Surgical Reporter, March 26, 1892.

Serious objections were raised against it by very high authorities. "Flügge, Baumgarten and Weigert, in particular, have opposed Metschnikoff's views, stating that a reception of bacteria by the cells of the body only takes place when the former have already been killed, or at least, have been deprived of much of their vital energy by other influences. The phagocytes, they maintained, did not form an active and dangerous weapon of defense for the organism, did not stand in the foremost rows of the battle against the invading parasite, but were open graves behind the line destined to receive the fallen enemies or any other lifeless bodies or substances. Nothing, they said, compelled us to believe that these cells possessed a peculiar devouring and digesting power; they were, on the contrary, nothing but the buriers of the dead, removers of decaying matter. They maintained, further, that whenever healthy vigorous bacteria entered the cells these latter always fell a sacrifice to them and were irretrievably lost."* Later experimental investigations by Metschnikoff have, however, completely refuted the claim that leucocytes attack none but dead bacteria. He has proven by a series of admirably conducted experimental investigations, which are, especially, remarkable for the careful manipulative dexterity he practiced, that anthrax bacilli which have been extracted from leucocytes that had englobed them, will grow and multiply themselves when they are transplanted on suitable sterilized culture media, and, furthermore, that anthrax cultures obtained from bacilli that had been englobed by leucocytes, have lost none of their virulence; when bacilli from these cultures are introduced into the susceptible organism they act with the same vigor and

---

*Frankel's Text and Book of Bacteriology.

rapidity that characterize anthrax bacilli which have not been thus treated. Now the advocates of phagocytic immunity attach great importance to the fact that leucocytes attack and destroy virulent bacteria, and its value would not be questioned if there were not important exceptions to this behavior of leucocytes. In truth, the Metschnikovian theory would be invulnerable if the leucocytes invariably made war upon invading microbes, but unfortunately, this is not the case. Leucocytes do not always rush upon invading bacteria, on the contrary, they as frequently rush the other way. Nor do they always protect the organism from infection by destroying the invading enemy; they as frequently retreat without even offering battle and leave the road to infection wide open.

The following statement embraces the accepted facts relating to the behavior of leucocytes towards infectious bacteria, viz.:

1. When virulent bacteria are inoculated into an animal which had previously been made immune from this bacterium, the white-cell elements of the animal will quickly migrate to the place of invasion, and will devour and assimilate the invading microbes. But when the same bacteria are introduced into an animal of the same species which has not been made immune, its leucocytes will not migrate to the place of invasion, and they will not attack or destroy the microbes; on the contrary, the leucocytes are repelled by the microbes, and the animal becomes infected.

2. A different result is obtained if the virulent bacteria are attenuated before they are introduced into the body, and, if the number first introduced is small, and then gradually increased in successive inoculations. Under these circumstances the animal becomes immune from this microbe, and

amounts of the virulent bacteria that would be fatal to a susceptible animal, can be safely introduced. The leucocytes will now destroy the microbes which would have destroyed the animal before its immunization.

Before the phagocytic theory of immunity, which regards the phagocytes as the defenders of the organism from invasions of pathogenic bacteria, can be accepted as proven, it must therefore explain the difference in behavior of the leucocytes of an immune and a susceptible animal towards virulent bacteria, and also the rationale of immunization which is acquired by means of attenuated forms of virulent bacteria. This is fully recognized by the advocates of this theory and they attempt to meet it by claiming that the leucocytes of a susceptible animal acquire the ability of successfully resisting virulent bacteria by attacking first the weakened (attenuated) forms; it is claimed that the leucocytes, by a process of training, become accustomed to, and are then able to destroy, the virulent forms of the bacteria. For example, when infectious bacteria, which repel and paralyze the leucocytes of a non-immune animal, have this power weakened by attenuation, and are then introduced into the body, in at first small and then in gradually increasing doses, the leucocytes become accustomed to these bacteria and, from successful encounters with the attenuated forms, the leucocytes gradually acquire, by a process of training or education, the requisite skill or power to destroy the virulent forms of these microbes. But the education or training of leucocytes is not sufficient, of itself, to explain the phenomena of immunity, for the life of the leucocyte is of brief duration when compared with the term of acquired immunity. It therefore became necessary to assume, that the education which leucocytes acquire from successful encounters with attenuated microbes is transmitted to their progeny as qualities of inheritance. And, it must be assumed, furthermore, that a special training of the leucocytes must occur, from encounters with specific bacteria of every infectious disease to which the individual has been subjected, and, that the educational advantages acquired by the leucocytes from such encounters, with different varieties of infectious bacteria, are all transmitted as qualities of inheritance.

The improbability that cells as simple in structure as leucocytes can be educated in the manner described, and that qualities thus acquired, involving as they must mental phenomena of no mean order, are transmitted by inheritance from parent to progeny through indefinite generations of leucocytes, is not the only difficulty which besets this hypothesis. Its inability to explain how immunity is acquired by inoculations of the sterilized products of bacteria, is fatal in its consequences, and has forced this hypothesis into a change of base. No bacteria are used in this method of inducing immunity; none are introduced into the body, and, consequently, the leucocytes have had no opportunity of acquiring immunity from virulent bacteria by encounters with attenuated forms of such microbes, yet the immunity which results from this method, is equally as good as that induced by inoculations of attenuated microbes. To meet this objection, it is claimed that it is the products of the bacteria, and not the bacteria themselves, to which the leucocytes become accustomed ; when these poisonous products are introduced into the organism, beginning with small doses, and gradually increasing the dose with each successive inoculation, that the leucocytes become gradually accustomed to the poison and eventually immune from toxic doses of it. Immunity acquired by the habitues of opium, alcohol and tobacco are cited as an analogous process; it is very questionable, however, whether immunity from toxic substances like alcohol, opium or tobacco is analogous to that which leucocytes acquire from bacterial poison. Immunity from alcohol and other poisons is terminated by a discontinuance of the drug, or, at most, with the individual's life; it is never transmitted to his posterity as a quality of inheritance. On the other hand, the immunity which leucocytes acquire from bacterial poisons does not terminate when the poison is withdrawn, or even with the life of the leucocite, but is thought to be transmitted by inheritance through many generations of these cells. It may be further urged against the hypothesis, that the immunity of leucocytes from bacterial poisons which is gradually acquired by accustoming these cells to the poison, does not explain how immunity is acquired by a single attack of an infectious malady; in this case the poison is not gradually introduced in increasing doses, but in such large amounts that the organism is overpowered and the leucocytes paralyzed, yet, the individual, if he recovers, will thereafter be immune from this disease. In this case the leucocytes are overwhelmed and paralyzed by the bacterial poisons; no educational or training opportunities are given them, and, consequently, the resulting immunity must depend on other causes than trained leucocytes. To avoid these and other difficulties which beset it, the phagocytic theory has finally intrenched itself behind the breast-works of chemotoxis.

"A very curious vital phenomena which has long been known in certain uni-cellular organisms—such as the fresh water amœba and in the leucocytes of both the cold and warm blooded animals—is their response by movement to contact with solid substances. Thus the amæba floating free in fluids tends to assume a spheroidal form and to remain immobile. When, however, under suitable conditions, it touches a solid surface, like that of a glass slide, it sends out pseudopodia and performs those curious progressive evolutions known as the amœboid movement. Essentially the same series of movements is observed in leucocytes when they, under favorable conditions, come in contact with solid surfaces—such as a glass slide or the walls of the body lymph spaces. This faculty in these primitive forms of life, consisting of a lump of protoplasm, is called *tactile sensibility*, and it is in virtue of this that many of the remarkable and useful evolutions of the leucocytes in the body transpire.''

"It was found by Pfeffer, a good while ago, that some of the lowly vegetable organisms endowed with locomotion, the *flagelletta bacteria*, etc., were capable of moving toward or away from certain substances which exerted a chemical action upon them.   This property he designated as *chemotoxis*, and further postulated as *positive chemotaxis* the attracting effect, and as *negative chemotoxis* the repelling effect, on such organisms of the chemical substances.

''The same condition of affairs is found to exist in the leucocytes of both warm and cold-blooded animals, and the conditions and bearings in them of this positive and negative chemotoxis were studied in detail by Massert and Bordet and by Grabritchevski, in 1890.   The latter observer has grouped, as the result of his experiments, certain chemical substances in accordance with their action in this way upon leucocytes. Thus, in the group of substances exciting a negative chemotoxis, we have concentrated salt solution, 10 per cent; lactic acid; quinine, 0.5 per cent; alcohol, 10 per cent; chloroform; jequirity; glycerine; bile. Substances having no effect—indifferent chemotoxis—are distilled water; aqueous humor; dilute salt solution, 0.1 to 1 per cent; carbolic acid, 1 per cent solution; antipyrine; glycogen; pep-

tone; beef-tea; blood.   Among the most prominent substances exciting a positive chemotoxis are especially steril·
ized and non-sterilized cultures of various pathogenic
bacteria.

"The general method of testing the powers of these various substances is to fill small capillary glass tubes, closed
at one end, and to thrust these beneath the skin of the animal.   After a few hours these tubes are withdrawn and
their contents examined.   Into the tubes filled with substances inciting positive chemotoxis the leucocytes crowd
in great numbers, while they are held away from the tubes
having negative chemotoxic contents, and when filled with
indifferent substances there is no effect at all."

Starting from these physiological facts of chemotoxis, the
phagocytists have constructed a theory, in which they claim
that the different behavior of the leucocytes of an immune
and non-immune organism toward pathogenic bacteria, is
caused by the chemotoxis of the bacterial products, and
that this influence is positive or negative according to the
degree of concentration of the poison, and its localization in
the body.   For illustration: the products of the causative
bacteria are assumed to remain permanently in the previously infected individual who is, therefore, immune from
this bacterium.   When, now, this bacterium is inoculated
into an organism which has been thus made immune, it is
assumed, the products which it secretes will exert a
positive chemotoxis on the leucocytes; that is, will attract
them to the point of invasion.   But when the bacterium is
introduced into a non-immune organism, which does not
contain the bacterial poison, the secreted products exert a

---

*Studies on the Action of Dead Bacteria in the Living body, by T.
Mitchell Prudden, New York Medical Journal, Bacteriological World.

negative chemotoxis upon the leucocytes; that is, will repel them, and the organism becomes infected.

A. A. Kauthack, F. R. C. S., in the recent discussion on "Phagocytosis and Immunity," at the London Pathological Society, has the following to say regarding this hypothesis:

"*Chemotoxis* has been of late introduced as a welcome support of the phagocytic theory. Certain chemical substances, according to their degree of concentration, attract or repel the phagocytes, but if the phagocytes are in a medium already containing these chemical substances, equally distributed, they will not be attracted on adding more of these substances, unless they be added in a highly concentrated condition.

"Analyzing this, we have three possible cases:

"(*a*) The metabolic products of the microbe are present in equal quantity in the tissues, and at the seat of lesion; therefore, no migration of phagocytes can take place.

"(*b*) If the degree of concentration is higher at the seat of lesion, we shall have a migration of leucocytes thereto.

"(*c*) If the degree of concentration is higher in the blood, the phagscytes may migrate back into the blood.

"This teaches us nothing as to the acquisition of immunity, for (1) if true, it is simply a statement of concomitant facts; (2) on such a view, our successes in the production of immunity become merely a matter of good fortune; (3) it supposes, without the slightest foundation, that in an animal, immunized against a microbe, the toxine of such microbe circulates in its vessels, and, lastly, it neglects the fact that, besides diapedesis, we have, as a rule, hand in hand with positive chemotoxis, a great general increase of phagocytes and leucocytes, depending of necessity on a stimulation, not of the phagocytes and leucocytes, but of the elements giving birth to these bodies.''

Hertzwig, in his monograph, sums up this theory in the following manner: "(*a*) In an immunized organism, the negative chemotoxis, which the body exhibited before being made immune, has been replaced by a chemotoxis, behaving positively towards the specific virus.'' If so, we may ask why and how?

We are also told that that effects of the metabolic products of microbes show themselves as positive and negative chemotropism, and also as acquired irritability (Reiznackwirkung). ''The existence of positive chemotropism explains the localization of the virus through the attraction of leucocytes and subsequent phagocytosis.'' The existence of negative chemotropism explains the possibility of general infection of the body by means of a diffusion of the micro-organisms.

By means of suitable injections of bacillary products, a negative may be changed into a positive chemotoxis, and thus a cure be effected.

But, aside from the mysticism of this hypothesis, and the *prima facie* evidence it bears of being a scientific absurdity, Buchner and his associates in Munich have concluded, from numerous experimental investigations, that the products of bacterial growth are only nerve poisons, with but little or no powers of inducing emigration of leucocytes, whereas the bacteri-protean, the plasma contents of the bacteria cell, itself, possesses this power in a high degree.

''It has been repeatedly shown by numerous observers that sterilized cultures of various pyogenic bacteria—such as *staphylococcus pyogenes aureus, bacillus pyocyaneus,* etc.— were as capable of producing suppuration as were the fresh living cultures. But it was believed that this was due to

the retention of a toxic substance furnished by the life processes of the germ which had not been destroyed by the
sterilization, but clung about the dead germ bodies. Although Wyosokovitsch had filtered off the fluid from sterilized anthrax cultures and found that the filtrate was not
pyogenic, while the solid material was, he inferred only
that the toxic material assumed to cause suppuration was
not soluble in the nutrient fluid.

"Buchner had also shown in the course of some other
experiments, that the sterilized emulsion of the so-called
pneumo-bacillus of Friedlander, subcutaneusly injected,
could cause suppuration in rabbits and guinea-pigs. He
found further that if such a sterilized emulsion were allowed to stand for some time, so that the solid could be
separated from the fluid parts of the mass; the fluid part did
not cause suppuration, while the solid part did. That the
effect of such sterilized bacterial emulsions was not due to
their mechanical effects in the tissues was shown by such
control experiments as the introduction of powdered char-
coal, infusorial earth, magnesia, potato emulsion, etc., beneath the skin, with negative results.

"By a series of manipulations similar to that practiced
with the pneumo-bacillus, Buchner now tested the effect of
sterilized emulsions of cultures of seventeen different species of bacteria, among which may be mentioned *staphylococcus pyogenes aureus, staphylococcus cereus flavus, sarcina
aurantiaca, bacillus prodigiosus, bacillus fitzianus, bacillus
cyanogenus, bacillus megatherium, bacillus subtilis, bacillus
coli communis, bacillus acidi lactici, bacillus anthracis, proteus
vulgaris, Finkler's comma bacillus, etc.* The injection of one
cubic centimetre of the sterilized emulsions of each of these
germs resulted within two or three days in an aseptic—that
is bacteria-free—purulent infiltration in the subcutaneous

tissue at the seat of injection. Ou the other hand, the clear fluid obtained by sedimentation from the sterilized emulsions of *bacillus cyanogenus, bacillus megatherium* and *bacillus anthracis* induced no suppuration while the separated sediment invariably did.

"But more definite proof of the importance of the bacterio-protein in inducing suppuration was still needed and Buchner proceeded to separate it from cultures of the pneumo-bacillus after the method of Necki, by digestion of masses of culture in dilute potash and precipitation with acetic or hydrochloric acid. The precipitate separated by filtration was again dissolved in dilute potash solution and reprecipitated. This was done the third time, and at last the purified product was brought into solution. This material gave the chemical reaction of an albuminoid body. Subcutaneous injection of this material in rabbits in some cases was followed by a gathering of leucocytes, in others not. As it seemed likely that on simple subcutaneous injection the material was readily and rapidly absorbed before it produced local effects, recourse was had to a method of experiment used by Councilman in his well-known studies on suppuration. Small glass tubes, drawn out at the euds, were filled with pneumo-bacillus protein, sealed up, and sterilized by steam for an hour. These were then introduced, with strict antiseptic precautions, beneath the skin of rabbits, shoved away from the opening, and, after they were healed in, their tips were broken off. After five days the tubes were exposed. Around the openings of these, as well as extending deep into their interior, were masses and plugs of leucocytes. Cultures showed no living bacteria. Control experiments with tubes filled with salt solution showed no collection of leucocytes.

"The next thing to be done was to carry on a similar

series of experiments with other well-known pathogenic bacteria. To this task Buchner and his associates addressed themselves in a series of studies as yet not fully published. But, even so far as their results are known, some most significant facts have been elicited. Buchner endeavored to separate by the method of Necki (see above) the bacterio-protein from about fifteen species of bacteria, but in many of these the attempt was unsuccessful, because sufficient solution and extraction of the proteid ingredients of the germs did not occur. The *bacillus pyocyaneus* gave the most abundant albuminous extract, but a sufficient amount was obtained from *staphylococcus pyogenese aureus, bacillus typhosus, bacillus subtilis, bacillus acidi lactici*, and from the red potato bacillus for animal experiment. It was, in fact, found that capillary tubes filled with the purified proteids from all these species of bacteria and placed beneath the skin of the rabbit showed after two or three days, extending into the open end, a plug of fibrinous pus several millimetres in length. This plug was found, on microscopical examination, to consist largely of leucocytes.

"That the ordinary chemical decomposition products of bacterial cell life are not concerned in inducing this positive chemotoxis in the leucocytes was shown by introducing beneath the skin of rabbits tubes filled with such substances as butyrate and valerianate of ammonia, trimethylamin, ammonia, glycocoll, leucin, tyrosin, urea, etc. These were, for the most part, wholly without effect upon the leucocytes, only glycocoll and leucin exciting in some cases a moderate chemotoxis, not at all to be compared, however, with that of bacterio-proteins.

"Now, what attracts the leucocytes into the vicinity of a particle of dead and useless muscle, or cartilage, or connective tissue which they are to absorb and remove? Certainly not bacterial poison, certainly not bacterial proteids; for with what may be called the normal phagocytic functions of the leucocytes bacteria having nothing to do. Having shown that a proteid substance derived from the bacterial cells was capable through chemotoxis of attracting leucocytes, Buchner now studied in a similar way the effects of closely allied substances—namely, the so-called vegetable caseins, gluten casein from wheat, and legumin from peas, both separated by precipitation from alkaline solutions. Both of these substances were capable of exciting the most marked chemotoxis in the leucocytes of rabbits. Moreover, as it has been shown that vegetable casein exists as such in the grain of cereals and of the leguminosæ, he introduced beneath the skin of rabbits or guinea pigs, under strict antiseptic precautions, masses of wheat and pea meal, and found that within two days these masses were surrounded and penetrated by enormous masses of leucocytes. Cultures from these masses proved the entire absence of bacteria. Starch introduced subcutaneously under the same conditions induced no gathering of leucocytes.

"But still another step remained to be taken. As the gathering of the leucocytes about dead organic fragments in the tissues which are to be removed, as so often happens, cannot be ordinarily due to bacteria or bacterio-protein, so, also, interesting as the observations may be, can vegetable proteins have no part in the matter. So alkali albumenates were prepared and purified, in a manner similar to that employed with the bacterial and other vegetable proteins from muscle, liver, lungs and kidneys of rabbits. These tested in the same way were all found to strongly attract leucocytes when introduced beneath the skin in the tubes. Of the alkali albumenates prepared from blood, fibrin, yolk and white of egg, only the blood and yolk of egg showed moderate power of exciting positive chemotoxis.

"These experiments show that it is only certain of the decomposition products of animal tissue which possess chemotoctic powers, and that these, as a rule, are the earlier and not the ultimate products of the decomposition.

"Finally, as it has been shown that a general leucocytosis is apt to be associated with febrile inflammatory processes, Buchner and Roemer studied the effects of intravenous injections in rabbits of these various chemotoctic proteids. They found that within eight hours of their introduction into the blood there was a marked leucocytosis lasting for several hours, and that this might be heightened by repeated injections. Thus they found, by a daily injection of 2 cc. of an eight per cent solution of the bacterio-proteins of *bacillus pyocyaneus*, the relation of white to red blood cells, which at first was 1 to 318, was on the second day 1 to 126; on the third day, 1 to 102; on the fourth morning, 1 to 73; and on the same evening, 1 to 38. From this time on no increase was noted. The absolute number of the red blood cells remained unchanged, while there was an absolute sevenfold increase in the number of leucocytes. Gluten casein, as well as alkali albumenate from muscle, injected into the blood, showed similar but less pronounced effects."*

The ease and clearness with which the physical theory philosophically explains the phenomena of fermentation, infection and immunity, is certainly significant of its truth, and in no instance is this more effectively shown than in its explanation of the different behavior of leucocytes of the immune and non-immune organism towards infectious bacteria. This problem which, we have seen, can not be satis-

*T. Mitchell Prudden, ibid.

factorily explained from the standpoint of phagocytic immunity becomes, however, a very simple matter when investigated from that of the physical theory. The first, and
an important difference between the two theories is that relating to the rationale or modus of formation of ptomaines
or bacterial products. The phagocytic theory regards these
as secretions or excretions of infectious bacteria ; the
physical theory believes they are derived from albuminoid
molecules of the body which have been disrupted, or
changed in molecular structure by the wave impacts—
dynamic energy—of virulent bacteria. The first theory
requires but one factor—the pathogenic bacteria; the second theory requires two factors—the pathegenic bacteria
and susceptible albuminoid molecules. The second difference between the two theories is that which relates to the
immediate cause of immunity. The phagocytic theory
places this in the behavior of the leucocytes of the organism ; it is believed these cells defend the organism
against bacterial invasions by means of inherent or acquired
qualities, and that these latter are transmissible from parent
to progeny. The physical theory places the immunity of
the organism in the molecular structure of its albuminoids;
when these produce waves that move in unison with those
produced by a pathogenic bacterium they are susceptible
albuminoids, and can be disrupted and converted into bacterial products—toxalbumins—by this microbe. But, if the
waves of albuminoid molecules do not move in unison with
those of the pathogenic bacterium, then such albuminoids
are immune from this microbe; it can not shake apart and
convert them into poisonous albumins. ·

When, now, pathogenic microbes are introduced into a
non-immune organism, they immediately proceed to shake
apart, by their dynamic energy, the albuminoid mole-

cules whose waves vibrate in the same periods as those of the bacterium, and the molecules liberated by this action at once recombine into poisonous albumins, which paralyze or drive away the leucocytes and infect the body. But, when this same microbe is introduced into an immune animal we may expect a different state of affairs, the albuminoid molecules of this animal do not move in unison with those of the bacterium (the cause of this has already been explained), therefore, they are not disrupted by it, and no poisonous albumins are formed which paralyze or drive away the leucocytes. On the contrary, they flock to the locality, from the law of physiological leucocytosis, and speedily devour the microbes.

Having now presented the features and claims of phagocytosis we will determine what explanation it can give of the phenomena of infection and immunity. Beginning with the phenomena of infection we have first, the incubative period of infectious diseases; second, the varying degree of virulency of pathogenic bacteria, and third, the self-limited duration of acute infectious diseases.

1. As the incubative period is that which begins with the introduction of pathogenic bacteria and ends with the beginning of pathological symptoms, explanation should be made why the symptoms are delayed, often for several days, after the body is infected. The assumed war between the leucocytes and bacteria begins immediately after the introduction into the body of these invading organisms; if, therefore, the pathological symptoms are due to this warfare they should at once become manifested.

2. While the phagocytic theory may recognize that this degree of virulency of bacteria corresponds to that of malignancy of the resulting disease, it can not, as a theory,

explain the cause of this varying degree of virulency of bacteria.

3. Why do acute infectious diseases run a typical course and terminate from the action of their own laws? What explanation can phagocytosis make of this well known phenomenon? For example, why, when the microbes of small-pox or of scarlet fever were apparently in full possession, should subsidence of the disease occur? If phagocytosis offers a philosopnical explanation of this phenomenon I have failed to find it.

Immunity may be a quality of inheritance—natural immunity,— or an acquired quality—acquired immunity. Regarding the phenomena of natural immunity the phagocytic theory can give no philosophical explanation. To say that the leucocytes of an animal that is naturally immune from an infection, have those combative qualities which, in less favored animals, can be acquired by artificial means only, without giving a philosophical explanation of the cause of the differing qualities of the leucocytes in the two cases, can not surely be accepted as a philosophical explanation of the phenomena; it must be regarded simply as a statement without explanation. This theory is next called upon to explain how immunity is acquired by a single attack. It will not do in this case, as already shown, to say that the leucocytes have acquired the requisite ability by previous training; that the cowardly leucocytes of a susceptible animal have been converted into valiant warriors of the immune, by first pitting them against a few of the foe, or by turning them loose on the hospital camp—as it were—and allowing them to devour the sick and feeble of the invading microbes. In this case the invading army attack in great numbers; they drive away or paralyze the defending leucocytes, and yet, if the individual recovers, he will thereafter be immune from other invasions of this microbe.

It is claimed that phagocytic immunity is an impossibility in the exanthemeta, a class of diseases from which immunity more frequently follows a single attack than any other. Whether this claim, in its general application, can be sustained may admit of doubt, but in its application to erysipelas, one of this class of diseases, there can be no reasonable doubt of its truth. The invading microbes of this disease are limited in their extent of action to the skin which is a territory that phagocytes can not invade; it is evident, therefore, that the immunity which follows a single attack of erysipelas must result from some other agency than that indicated by the phagocytic theory.

In this connection it is proper to inquire what philosophical explanation is offered by this theory why immunity from infectious bacteria is so variable in its degree and duration? The facts are, that acquired immunity from some infectious diseases continues during the life of the individual; from other infectious diseases its duration is much shorter, whilst other diseases, whose infectious nature is undoubted, may recur in the same individual almost without limit. One attack, in these cases, gives the individual no protection whatever from subsequent attacks of the same disease. If the cause of immunity is to be found in trained leucocytes, why do the combative qualities which these cells are thought to acquire persist in some cases, and quickly disappear in others? And why—as similar causes produce similar results—do not leucocytes, in their successful battles against invading microbes acquire, in all cases alike, the requisite skill to successfully resist future invasions of the microbe?

IMMUNITY ACQUIRED BY MEANS OF ATTENUATED BAC-TERIA.

This form of immunity furnishes phagocytosis its strongest support, for it has been established by actual demonstration that leucocytes will take living attenuated forms of virulent bacteria into their bodies and therein digest them. Now, it is well known that inoculating a susceptible animal with attenuated forms of pathogenic bacteria will confer immunity from the virulent forms of the microbe, and it has been established by Metschnikoff and others, that the leucocytes of an animal which has been made immune will englobe and digest virulent microbes; it is, therefore, assumed from these established facts, that the behavior of the leucocytes of immune animals toward the virulent microbes has been acquired by training, and that the immunity enjoyed by the animal is the result of phagocytic action.

IMMUNITY ACRUIRED BY THE PRODUCTS OF BACTERIA.

When the various methods by which immunity is artificially produced are critically examined and compared, I think it will be seen that all these methods, for example by a single attack; by infectious bacteria; by attenuated bacteria, and by the products of infectious bacteria, will practically be resolved into the same thing; and that in each and every case it will be found that the products of infectious bacteria are the immunizing agents.

If, now, the several different methods of producing artificial immunity are resolvable into the single one—by means of the bacterial products—then the phagocytic theory must prove itself to be in line with this method;—or, upon failure to do this, it should be retired from the list.

It has been stated already that the claim first advanced

by this theory, that leucocytes acquire the requisite ability to destroy virulent bacteria through successful encounters with the attenuated forms of these micro-organisms, is totally inadequate to explain how immunity is acquired by bacterial products. As no bacteria, whatever, are introduced into the body by this method, some other explanation than that of trained leucocytes had to be made. That which was adopted, and the one which, in a measure, is supported by analogies furnished by physiological chemotoxis, and the accustoming of individuals to drugs, claims that it is the poisonous products of bacteria, and not the bacteria themselves against which leucocytes acquire immunity by means of small amounts of these poisons being gradually introduced into the body of a susceptible animal. Aside from the reasons given why the analogy which is claimed to exist between the following processes, viz: that by which leucocytes become accustomed to, and relatively immune from bacterial products, and that by which individuals become accustomed to and relatively immune from poisonous drugs, such as alcohol, opium and tobacco, there are other and serious objections to this hypothesis. For example, if accustoming the leucocytes to bacterial products is the cause of acquired immunity, then an individual who is immune from a bacterium would, in the same degree, be immune from its poisonous products. The fact that this is not the case,—that immunity from a bacterium does not confer the same degree of immunity from its products, is strongly inimical to this hypothesis.

In view of the fact established by Hueppe and Wood, "that a species of bacteria, clearly distinct from anthrax bacillus, apparently innocuous, and strictly saprophytic, was able to secure even very susceptible animals, such as mice and guinea pigs, against anthrax;" and the further fact announced by Wooldridge, that he obtained immunity against anthrax by means of a substance which has no connection with the vital process of bacteria; and the still further fact, that immunity from several diseases has been conferred by inoculating one kind of bacteria (all of which strongly militate against the assumed hypothesis that acquired immunity of the individual has resulted from his having become accustomed to the products of the associated bacterium) we must seek the causes of acquired immunity in other changes of the body, deeper and more profound, than those supposed to occur in its leucocytes.

CHAPTER VIII.

THE PHYSICAL THEORY.

The following interesting description of Zadig's method, from the pen of his biographer, Aronet de Voltaire, was used by Prof. Huxley as an introduction to an article on "Retrospective Prophecy as a Function of Science,"* and is reproduced here, as it serves to illustrate the inductive method,—reasoning from facts, phenomena and analogies back to the ultimate causes of motion in matter embraced in that branch of mechanics called kinematics,—by which the physical theory, after considerable time and thought, was eventually worked out:

"One day, walking near a little wood, he saw, hastening that way, one of the queen's chief eunuchs followed by a troop of officials, who appeared to be in the greatest anxiety, running hither and thither like men distraught, in search of some lost treasure.

"'Young man,' cried the eunuch, 'have you seen the queen's dog?' Zadig answered modestly, 'A bitch, I think, not a dog.' 'Quite right,' replied the eunuch; and Zadig continued: 'A very small spaniel who has lately had puppies; she limps with the left fore leg, and has very long ears.' 'Ah, you have seen her, then?' 'No,' answered Zadig, 'I have not seen her; and I really was not aware that the queen possessed a spaniel.'

---

*Popular Science Monthly, August 1, 1880.

"By an odd coincidence, at the very same time, the hand-
some horse in the king's stables broke away from his
groom in the Babylonian plains.   The grand huntsman and
all his staff were seeking the horse with as much anxiety as
the eunuch and his people the spaniel; and the grand huntsman asked Zadig if he had not seen the king's horse go that
way.

" 'A first rate galloper, small hoofed, five feet high; tail
three feet and a half long; cheek pieces of the bit of twenty-
three carat gold; shoes silver?' said Zadig.

" ' Which way did he go?' 'Where is he?' cried the
grand huntsman.

" ' I have not seen anything of the horse, and never heard
of him before,' replied Zadig.

"The grand huntsman and the chief eunuch made sure
that Zadig had stolen both the king's horse and the queen's
spaniel, so they haled him before the high court of Dester-
ham, which at once condemned him to the knout and trans-
portation for life to Siberia.   But the sentence was hardly
pronounced when the lost horse and spaniel were found.
So the judges were under the painful necessity of reconsid-
ering their decision; but they fined Zadig four hundred
ounces of gold for saying that he had seen that which he
had not seen.

" The first thing was to pay the fine; afterward Zadig
was permitted to open his defense to the court, which he
did in the following terms:

" 'Stars of justice, abysses of knowledge, mirrors of truth,
whose gravity is as that of lead, whose inflexibility is as
that of iron, who rival the diamond in clearness, and possess no little affinity with gold, since I am permitted to address your august assembly, I swear by Ormuzd that I have

never seen the respectable lady dog of the queen, nor beheld the sacrosanct horse of the king of kings.

" 'This is what happened: I was taking a walk toward the little wood near which I subsequently had the honor to meet the venerable chief eunuch and the most illustrious grand huntsman. I noticed the track of an animal in the sand, and it was easy to see that it was that of a small dog. Long faint streaks upon the little elevations of sand between the foot marks convinced me that it was a she dog, with pendent dugs,—showing that she must have puppies not many days since. Other scrapings of the sand, which always lay close to the marks of the fore-paws, indicated that she had very long ears; and, as the imprint of one foot was always fainter than those of the other three, I judged that the lady dog of our august queen was, if I may venture to say so, a little lame.

" 'With respect to the horse of the king of kings, permit me to observe that wandering through the paths which traverse the wood, I noticed the marks of horseshoes. They were all equi distant. "Ah!" said I, "this is a famous galloper." In a narrow alley only seven feet wide, the dust was a little disturbed at three feet and a half from the middle of the path. "This horse," said I to myself, "had a tail three feet and a half long, and, lashing it from one side to the other, he has swept away the dust." Branches of trees met overhead at the height of five feet, and under them I saw newly fallen leaves; so I knew the horse had brushed some of the branches, and was therefore five feet high. As to his bit, it must have been twenty-three carat gold, for he had rubbed it against a stone, which turned out to be touch-stone, with the properties of which I am familiar by experiment. Lastly, by the marks which his

shoes left upon pebbles of another kind, I was led to think
that his shoes were of fine silver.'

"All the judges admired Zadig's profound and subtle dis-
cernment, and the fame of it reached even the king and
queen. From the anterooms to the presence-chambers,
Zadig's name was in everybody's mouth; and although
many of the magi were of the opinion that he ought to be
burned as a sorcerer, the king commanded that the four
hundred ounces of gold which he had been fined should be
restored to him. So the officers of the court went in state
with the four hundred ounces; only they retained three
hundred and ninety-eight for legal expenses, and their ser-
vants expected fees.''

From the beginning of these investigations we were
strongly impressed with the striking analogies between
the phenomena of fermentation and infection which have
long been recognized by medical writers and observers, and
upon which is based that classification of diseases wherein
those believed to be contagious or infectious are classed as
zymotic or fermentation diseases. A century ago one of the
pioneers of chemistry made the memorable statement, "He
that understands the nature of ferments and fermentation
shall probably be much better able than he that ignores
them to give a fair account of divers phenomena of certain
diseases (as well fevers as others) which will perhaps never
be understood without an insight into the doctrine of fermentation.'' Further investigation into this subject not
only confirmed me in my first belief, but convinced me that
such remarkable analogies could not occur unless the underlying principles which determine these phenomena are the
same for both processes, or, at least, are so nearly related
that the philosophy of one process cannot be known without understanding that of the other. At the same time,

when I sought an explanation of the established facts of fermentation from existing theories, it was found that these are conflicting and inadequate to give a philosophical explanation of the phenomena; that, in fact, the intimate nature of fermentation is as great a mystery as that of infection, and that an attempt to construct a theory of infection that would harmonize and explain the phenomena common to both processes cannot succeed under the present views of fermentation. Continued and thoughtful consideration of the subject of fermentation in all its different aspects, and from the standpoint of its different theories, finally led me to seek for the unknown cause in those movements of the ultimate parts of matter, called atoms and molecules, which constitute the dynamics of chemistry, physics and biology. It was soon found that the laws of matter and motion, legitimately applied and interpreted, were able to explain all the phenomena of fermentation, and to harmonize and explain all the phenomena of infection.

The new theory of fermentation, which in part was evolved from investigations along new lines of thought, and in part derived from an old theory, I have concluded to call, for the want of a better name, the physical theory, although it rests as much upon the principles of chemistry and biology as it does upon those of physics. The philosophical explanation of phenomena which this theory gives is not confined to those of fermentation, it is my belief that it will furnish an equally lucid and rational explanation of the phenomena of infection and immunity. But confining these remarks, for the present, to fermentation, I will pass in brief review the known facts of this process, which have already been fully discussed in the first part of this essay, and will then state the accepted laws of motion in matter upon which I believe this process depends, and later on,

will give reasons for believing that the phenomena of
infection and immunity result from the same underlying
principles. I particularly invite your thoughtful attention to the facts and principles which are herein reviewed,
as it is upon these that I base the new theory of fermentation, and believe when these matters are clearly understood, and the modus of these agencies in producing fermentation is fully comprehended, that there will be no difficulty in understanding the explanation how these same
laws of matter and motion, as they are manifested in the
protoplasm of bacteria cells and the albuminoids of the
body, give rise to, and immunity from, infection.

Fermentation is the process of decomposition, or of conversion, effected by a ferment. (Foster.) A ferment is an
organized body, capable, in small quantities, of decomposing other organic bodies without yielding any of its own
substance to the product of the fermentation. (Foster.) It
has been stated that fermentation requires two factors, viz:
a ferment and a fermentable substance. Ferments are of
two classes, viz: living (organized) and non-living (unorganized) ferments. Both classes of ferments act alone by
contact, i. e., they cause a disruption and conversion of fermentable substances into other (new) substances—products
—without giving up any portion of their own substance.
Contact of the ferment and fermentable substance as a *sine
qua non* to fermentation is illustrated by the following experimental fact, viz: when a bladder filled with a sterilized
fermentable fluid is immersed in a similar fluid undergoing
fermentation, no change whatever occurs in the fluid which
the bladder contains; in other words, fermentation cannot
occur except the ferment is brought into immediate contact
with the fermentable substance. This one fact alone, it
would seem, is fatal to that theory which assumes that fer-

ment bacteria secrete unorganized ferments—enzymes—which are regarded as the active ferment agents. According to this hypothesis unorganized ferments, being soluble, would not be confined by the bladder membrane, but would readily pass through this by osmosis, and fermentation would occur in the grape juice inside this membraneous bag, as readily as outside it.

The products of fermentation are derived solely from the substance fermented; this substance is disrupted by contact with the ferment, and its molecules recombine to form these products. For example, alcohol, carbonic acid, etc., which constitute products of vinous fermentation, are derived from, and actually represent the fermentable substance—the sugar —which is decomposed by contact with the vinous ferment. A ferment does not indiscriminately decompose all fermentable substances; on the contrary, it is limited in its action to one, or, at most, to a few of these, and the products of its action are specific and distinctive of the causative ferment. The products of fermentation, have power to inhibit or arrest further action; so when these accumulate in certain proportional amounts in the fermenting liquids, they arrest the process and this cannot be re-established until the products are lessened or neutralized.

Living ferments are one-celled vegetable micro-organisms which, generally, belong to the class schizomocytes (bacteria). They grow and multiply as do other organisms and produce none but their own kind. When ferment bacteria are grown under unfavorable conditions of environment, such as unfavorable conditions of climate or of food supply, these organisms become weakened (attenuated) in their power of producing ferment products, and, if the attenuating agencies are pushed to an extreme limit, this power may be entirely destroyed. In some cases the change of biologic

habit produced in these micro-organisms by attenuation, becomes a natural habit that is transmitted by inheritance
through the race, and this, too, without changing the appearance, mode of growth, or habit in other respects of
such ferment organisms; e. g., the yeast ferment (saccharomyces cervisiæ) may thus be weakened in its functional
action, when it will produce but a small quantity of alcohol, etc., or even none at all without interfering with its
other properties, and this changed habit may be made
permanent and a quality of transmission.   Non-living (unorganized) ferments are that class which are formed in animal and vegetable cells. While these differ from the former
in not having procreative life, like the former they require
to be in contact with a fermentable substance before they
can disrupt this and convert it into ferment products.   The
products of these ferments are also specific and distinctive,
and their accumulation in the fermenting solutions will arrest the process, as in the other case, and these products,
like those of unorganized ferments, are derived solely from
fermentable substances; they add nothing from their own.

The striking similarity of phenomena of fermentation,
whether produced by one or the other class of ferments,
certainly indicates that the philosophy of action is the
same in both cases.   In fact, reflection upon the analogies of the two processes have led many observers to believe the intimate nature of fermentation is identical,
but his deduction is not borne out by existing theories; they cannot harmonize the phenomena and unify
the supposed causes. .Professor Sedgwick was almost
prophetic in his utterances when he said: "If, however, this theory be ever established, the final explanation
of fermentation must be sought in molecular physics."
Now, this is exactly where we have sought it, and it is in

molecular physics the physical theory finds the explanation that unifies the cause, and harmonizes all the phenomena of fermentation. But before entering into an explanation of this theory, we must briefly review some of the facts of molecular physics, which have elsewhere been fully elaborated, upon which this theory is based. They are as follows: The ultimate divisions of all material substance are its atoms and molecules. These bodies have motions in periodic time that are continuous, unvarying, and are distinctive of each kind of atom and molecule. When atoms and molecules combine to form simple or compound masses they do not lie in contact, but are separated by spaces filled by an elastic and highly attenuated medium called universal ether, luminiferous ether, or simple ether, which fills all space. The movements of atoms and molecules in periodic time produce wave motions in the ether which correspond in periods with those of the atoms and molecules. Ethereal-wave motions, like other wave motions, interfere with each other, e. g., waves of the same period increase, while waves of opposing periods decrease, or, it may be, destroy each other.

When, in the light furnished by these accepted truths, we picture to the mind the mechanism and movements of molecules, e. g., those of ferments, our attention will be fixed, first, on the grouping of atoms to form molecules, and of molecules to form other, more complex molecules, which occur in almost endless variety; and, second, on the rythmic and unceasing motions of the atoms, in characteristic periods of time that are distinctive of each kind of atom; and, third, on the waves which these motions produce in the ether. It would be seen that the interference of wave motion known to occur between waves of water, waves of sound, and waves of light, also takes place here;

atomic waves of one kind increase others of the same periods, and decrease, or, it may be, destroy those having opposite periods.

If our argument is, so far, sustained by the principles of molecular physics which have been stated, we can safely carry this line of investigation further, and reasoning by deduction from the law of interference arrive at the following legitimate conclusions, viz: the wave motions of a molecule differ from those of its atoms; the waves of a complex molecule have periods which differ from those of its constituent molecules, and the wave motions of molecules have periods that are distinctive of the structure, i. e., the atomic and molecular grouping of each kind of molecule. But, fortunately, we are not compelled to rely wholly upon deduction for this information. The revelations of the spectroscope have conclusively proven that ethereal wave motions of a complex substance have periods which differ from those of its constituent atoms and molecules.

If we credit the teachings of modern science regarding the nature of force, which is defined as the efficient cause of all physical phenomena, we must accept the statement that this is derived from the motions of atoms, of molecules, or of mass. Vewing, then, a ferment, for illustration the yeast cell, from this standpoint, we must regard it as a molecular machine whose kinetic energy, or power of doing work, is derived from its molecular structure; that the nature of its atoms, and the order in which its molecules are constructed and grouped, give to the cell its characteristic wave motions and these determine the kind of work the cell can perform. Likewise grape sugar, the fermentable substance of grape juice, must be regarded as a molecular substance, although more simple in its structure and less fixed in its chemical bond than the yeast cell. If, now, the ether waves produced by the molecules of grape sugar dissolved in grape juice, i. e., the molecules liberated from mass contact and free to act, and those of the yeast cell, have periods which coincide, or nearly coincide in their rhythm, it does not require prophetic vision to mentally see that the successive impact of waves of the yeast cell, perhaps many million times in a second, falling upon those of the sugar molecules, must increase the amplitude of these which, in being driven back upon the atomic constituents, would increase their swing and drive the atoms further and further apart until they would finally be driven beyond their chemical attractions and the sugar would be disrupted, i. e., resolved into its atomic condition. It is, however, an established law of chemistry that atoms thus conditioned will immediately recombine into other, less stable and simpler substances. This, in fact, is what happens in the case under consideration. The atoms of the sugar recombine to form alcohol, carbon dioxide, etc., which collectively contain the same atoms in the same proportions, though differently combined, that grape sugar contains.

The intimate nature of vinous fermentations, as set forth in the illustration just given, serves to explain what I believe to be the *modus operandi* of all fermentations. All ferments, according to the physical theory, are organic substances having an atomic and molecular structure that gives them power to drive apart the molecules of other organic—fermentable—substances when these are brought into solution and otherwise suitably conditioned. The difference in physical, chemical, and physiological properties of the products of fermentation are caused by difference in the molecular structure of the ferments on one side, and the fermentable substance on the other; in other words, there is a definite relationship existing between the ferment,

the fermentable substance, and the products of every fermentation, which is caused by the molecular structure of
the substances named.

We then claim for this theory that it unifies the intimate
nature of all fermentations. It explains why ferments require
contact with fermentable substances, and why fermentation
is limited by a filter that will not allow the ferment to pass.
It explains why ferments have specific action, and why the
ferment products are distinctive of each kind of fermentation and are derived solely from the fermentable substances.
It explains the philosophy of attenuation, and that attenuated ferment micro-organisms produce less, or, it may be,
no ferment products because their molecular structure has
been changed by the attenuating causes. (The method by
which this is brought about is forcibly illustrated in the interference of sonorous waves which occurs when one of two
tuning forks having equal periods is artificially changed in
its periods of vibration, to which we have previously referred. Before the change both vibrated in unison and gave
out a continuous musical note; after the change this no
longer occurred; one vibrated more rapidly than the other
and the resulting note was a rising and falling sound, most
distinct when the waves from one supplemented those from
the other, and least distinct, or absent when they antagonized or destroyed each other. In this example the work
done is the sound produced, which is more or less according
to the amount of wave interferences. In the other case, the
two tuning forks are represented by the ferment bacteria and
the fermentable substance, which, before attenuation have
equal periods, but after attenuation the periods of the ferment are artificially changed and they no longer coincide
with those of the fermentable substance, consequently the
work done, in this case converting the fermentable substance

into ferment products, as in the other, is more or less depending upon the amount of wave interference; more when the waves coincide and less, or, it may be, none when the waves of the ferment are antagonized or destroyed by those of the fermentable substance.) Finally, the principles of molecular physics on which this theory is founded, makes the following explanation why the products of fermentation inhibit the action of the ferments, and why an accumulation of the products will arrest the fermentation. If disruption of fermentable substances is caused by molecular bombardment of the ferments, the product which results from this disruption must form under the same influence, i. e., the nascent atoms and molecules must recombine to form other substances—ferment products—under the influence of the molecular waves of the ferment. Now, it is evident that no compound can form against this influence unless its periods do not coincide with those of the ferment, for the same influence which disrupted one substance would surely prevent the formation of another having equal periods. It is seen, then, that the compound which would form from the nascent atoms and molecules, would be that whose molecular waves least coincide with those of the ferment. And when we consider that waves of the same kind must either supplement or antagonize each other, and that products of destructive metabolism, like those of fermentation, are simpler and more stable than are those from which they are formed, it follows that wave motions of ferment products antagonize those of the ferments, and that when the products have accumulated in sufficient quantity, they will inhibit the action of the ferments and thus arrest the fermentation.

From whatever direction we investigate this subject ·the conclusion will be reached, that the intimate cause of fermentation rests in molecular physics. If we start our investigations from accepted principles of molecular physics, as in the present exposition of this subject, the phenomena of fermentation will receive a philosophical explanation. Or, if we pursue the opposite line of inquiry, and start our investigation from the phenomena, and seek their explanaation by reasoning deductively from effects back to cause—a method which we were compelled in a measure to rely on in thinking out the cause of fermentation—the principles of kinematics will be evolved as a necessary result. This was illustrated in two instances while working out the physical theory; it was early seen that atoms must have specific vibrations to account for specific actions of ferments. I may have had some knowledge of the fact that atoms had some such action, but it was indefinite; investigation of authorities, however, verified the fact and proved the truth of the deduction. In the other instance I found it necessary that molecular waves of a compound should differ from those of its elements, in order to explain differing manifestations of energy. Investigation again sustained the deduction.

It is significant that, with few exceptions, all theories of fermentation which have been advanced, recognize molecular motion as the efficient cause of this process. This is notably true of Stahl's theory, and also that of Liebig's, while the catalytic theory of Berzelius cannot receive an intelligible interpretation except on this hypothesis. As far back as the fourteenth century and long before the birth of molecular physics, Stahl, reasoning from the then known phenomena of fermentation, reached an explanation of this process which he placed in "molecular activities." But the limited knowledge had at that time of this branch of physics, made it impossible to verify any deductions which might have been arrived at relating to molecular movement occurring in distinctive periods of time, therefore Stahl's conception of molecular activities lacked that definiteness which modern science gives them. Some four hundred years later this theory was taken up in a modified form by Liebig and became the most popular and universally accepted theory of that time. The indefinite, haphazard, molecular movements of Stahl's theory became, under that of Liebig, motions of decay (*motor-decay*). He claimed that ferments were organic substances undergoing retrogressive metamorphoses, and that fermentation was induced by the ferments imparting their own motions of decay to other substances, and thus inducing in these the same retrogressive changes. Under the leadership of the brilliant and scholarly Liebig, this theory continued to dominate the scientific world until the discoveries of Pasteur in fermentation, which were epoch making in their results, gave it the *coup de grace* by proving that a very large proportion of ferments are living, growing cells which have nothing to do with molecular motions of decay. In the face of this evidence Liebig's conception of ferments and fermentation can no longer be maintained. But in the light of modern science, if we regard ferments as molecular bodies whose molecular motions in definite periods of time are those of organic life, then the objections of Pasteur which were fatal to Liebig's hypothesis, will not in the least affect this theory.

We have reviewed the subject of fermentation at this length, because the remarkable analogies between the phenomena and those of infection lead us to believe that the intimate causes of the phenomena are the same in both processes; and that a clear understanding of the *modus operandi* of these causes, already set forth, will enable the reader to better comprehend how the same agencies give rise to infectious diseases and immunity from the same. The analogies referred to, and the reason for our belief in the unity of causation of the phenomena, will be shown by passing in brief review the phenomena of infection. These can be studied best as they are manifested in infectious diseases; they are as follows, viz:

This class of diseases are those known as "catching," because they are communicated from the sick to the well; the infectious cause is known to be a particulate substance that is often conveyed long distances in infected goods, etc., without losing its infectious qualities.

Diseases of this class, in their transmission, give rise to their own kind only; e. g., small-pox, measles, or yellow fever, convey no other infection than that of the respective disease.

Acute infectious diseases are self-limited in duration; they run a course that is typical of the disease, and, as a rule one attack confers immunity to the infected, from other attacks of the same malady.

Epidemics of infectious diseases manifest a marked variability in their degree of virulence; one epidemic or a part of an epidemic may be characterized by extreme virulence, another epidemic, or another part of the same, may be much less virulent.

Bacteriological research has furnished a vast amount of evidence, which in many cases is irrefutable, in support of that theory of disease which teaches that pathogenic bacteria are causatively related to infectious diseases; these bacteria elaborate poisonous substances called ptomaines, toxines and tox-albumins which are distinctive of the pathogenic bacterium concerned, and are the efficient causes of infection. Pathogenic bacteria, like ferment bacteria, are microscopic-one-celled-organisms which have, in both cases, distinctive functional powers.

Fermentation requires the harmonious action of a ferment and a fermentable substance, in like manner, our physical theory explains pathogenesis as depending upon the harmonious action of two factors, a bacterium and albuminoids; when these two factors have that similarity of molecular structure which enable the ether waves of the bacterium to disrupt the albuminoids, recombination of the liberated molecules produce the poisonous tox-albumins and pathogenesis results.

The products of pathogenesis, like those of fermentation, are then distinctive of the two active factors; when these are changed the products will be changed, and, as the body albuminoids are represented by innumerable isomeric forms of albumin, and on the other hand, a similar diversity of molecular combinations is found in the protoplasm of bacteria, there must result a diversity of poisonous ptomaines, from different classes of bacteria acting on some of the many isomeric forms of albuminoids which normally exist in the animal body. And, as the poisonous products of bacterial action are the causes of infection, and each separate ptomaine, or group of ptomaines derived from the same bacterium, excite characteristic and distinct reaction, or disease, it will be seen that different types of infectious diseases arise from the diversity of molecular structure of, first, bacteria; and, second, of the albuminoids of the body.

This same hypothesis will explain why bacteria manifest distinctive functional power, and also why bacteria that are known to be pathogenic to certain classes of animals, are non-pathogenic to other classes. The whole matter is resolved into the question of susceptibility or non-susceptibility of the albuminoid molecules of the animal to the molecular bombardment of a given bacterium. A pathogenic bacterium, from this standpoint, is one whose molecular waves can disrupt albuminoids of the body which, in re-forming, produce ptomaines. If we call such albuminoids *susceptible*, then we should call those which resist this bombardment *immune albuminoids*. The causes which, I believe, have determined the natural susceptibility and immunity of albuminoids have been fully presented in the discussion of "natural immunity," and require no further notice.

Going back to the analogies which exist between the phenomena of fermentation and infection, there yet remain, from our point of view, two important examples for consideration; one is furnished by the typical and self-limited course of infectious diseases; the other by the varying degree of malignancy which epidemics of these diseases manifest. Bearing in mind the established fact that the products of bacterial action are the true causes of infection, it will be seen that the same principles of molecular physics which give ferment products the power of inhibiting ferment bacteria, will give pathogenic products the power of inhibiting pathogenic bacteria, and, in both processes alike, an accumulation of products will arrest the process.

But, it may be asked, how do the pathogenic products arrest the disease when we are required to believe that they are its cause? Admitting they inhibit the bacteria and thus prevent a further formation of products, does not explain why those already formed do not continue to excite specific disease as long as they are retained in the body, or, in case of their removal, why the bacteria cannot again form new products and, in this way, continue the disease. We have already stated that Chauveau's theory attempts to explain this by, first, assuming that pathogenic products are retained indefinitely in the body, like ferment products are retained in fermentation vats, and, second, that these poisonous products do not continue their specific action because the individual gets accustomed to their presence.   The physical theory offers, I trust, a more philosophical explanation, as follows.   It has been stated in the discussion of fermentation, why the wave motions of a ferment and a fermentable substance must recur in a relative order of time that those of the first can disrupt and convert those of the second into ferment products, likewise the wave motions of a bacterium and those of ultimate albuminoid molecules, must recur in equal periods that the bacterium can convert the albuminoid molecules into poisonous albumins; the products in both cases, it will be remembered, give off wave motions that interfere with those of the respective bacteria, and, in this way, inhibit their action.   Now, while this action may be final so far as it concerns fermentation, the conditions are quite different in the animal organism; aside from its organic life and its ability to neutralize and eliminate harmful substances, must be considered the great diversity in the molecular structure of its closely allied albumins (the isomeric forms of albuminoids), with which we are specially concerned. This diversity of molecular grouping of the body albumins is necessary, that they may nourish and be converted into the different tissues of the body.   It is certainly a reasonable supposition, that a dynamic change in any single group of these bodies, may induce change in other groups, and that products of such changes may in some cases be harmless, and in others harmful.   At least it is evident in the case in point, where the group of albuminoids have similar vibrations with those of the bacteria, that the product which inhibits the action of the bacteria would react on those albuminoids having similar molecular vibrations.   If, now, this reaction, instead of disrupting the albuminoid

molecule, only changed its molecular grouping (just as attenuating agencies change the molecular grouping of bacteria), a corresponding change in its wave motions would
result as a consequence, and, in this case, the albuminoid
would no longer be vulnerable to the molecular bombardment of the bacterium, i. e., it would become immune from
this bacterium, and the disease would not only be arrested,
but as long as the albuminoids retain this new group-form,
the individual would have immunity from that disease of
which this bacterium is an etiological factor.   The conversion of *susceptible* into *immune albuminoids*, in the manner
described, need not destroy the nutritive or harmless qualities of the substance, it may still remain a normal constituent
of the body if, as I assume, this is a normal condition of albuminoids in naturally immune animals.   The self-limited
feature of acute infectious diseases, can be simply explained
by this hypothesis without assuming, what is very improbable, a harmless retention of poisonous substances in the
body.

The rationale of immunity which has been induced by
varying processes, for example, immunity against anthrax
by tissue-fibrinogen (Wooldridge), or, by a species of bacteria, clearly distinct from anthrax bacillus, apparently innocuous and strictly saprophytic (Hueppe and Wood), and
immunity from several diseases induced by inoculations
with one kind of bacteria (Roux and Chamberland), which
have caused so much trouble to existing theories, receives
a clear and intelligible explanation when we regard the
susceptibility of animals to, and their immunity from infectious diseases, as depending upon susceptibility or immunity of certain albuminoids of their bodies to the dynamic effects of bacteria.   Any agent that can render susceptible albuminoids immune from a bacterium, must so change the

molecular structure of the albuminoids that they will no longer vibrate in unison with those of the bacterium. This agent, whether it be tissue-fibrinogen or a bacterium, must of course have a molecular structure that corresponds in certain ways with that of the albuminoids, its molecular waves must be able to exert the requisite influence on those of the albuminoids before they can produce a change of molecular structure in these bodies. While the relation between the wave motions, in periodic time, which is required to exist between an immunizing agent and a susceptible albuminoid most frequently happens between the products of bacterial action on the one hand, and the related bacteria and albuminoid on the other, for reasons which have been given, it is not improbable that this happens less frequently between other substances, e. g., in those cases referred to; nor is it improbable that the change in the molecular structure of an albuminoid which a bacterium causes, may be such that will render the albuminoid immune from other bacteria, or it may be, susceptible to bacteria from which it had been previously immune.

If the change which is produced in susceptible albuminoids by the ptomaines is one of disruption, instead of one of molecular rearrangement, then a recombination of the molecules of the disrupted substances would form inhibitory substances that would antagonize the ptomaines; in other words, these substances, by neutralizing the infectious poison—ptomaine—would possess curative properties, and, like defensive proteids, when injected into an infected animal, may arrest the infection. Or, some albuminoids may be broken while others are simply changed, more or less, in their structure by one or the other ptomaine which is produced,—analogy and observation both teach that often more than one ptomaine are produced by the same bacte-

rium.   In this case the "defensive proteid" is not the only
immunizing agent, and its elimination from the body would
not destroy the immunity which had been secured by the
ptomaine.   The question whether the substances formed
from broken albuminoids, which, I believe, are those
called "defensive proteids," are retained or eliminated, will
depend on whether they are harmful or useful to the organism.   As a matter of observation, in many cases they are
eliminated— through the urine or milk in those cases
named—but on the other hand, it would seem, from the experiment of Hankin, that these substances are formed by
natural methods; in the naturally immune white rat, at
least, they become normal albuminoids of the organism.

If, now, immunity is secured in this way, we must explain why it is so variable in degree and duration.   The
answer which the physical theory makes to this, is derived by analogy from the phenomena of attenuated bacteria, and a conception how the dynamic forces of molecular physics produce these phenomena.   For illustration,
attenuation of bacteria consists in changing the molecular
structure of the protoplasmic bodies of these micro-organisms without producing visible change in their form, appearance, mode of growth or reproduction, and, furthermore, this molecular change— attenuation—is variable
both in degree and duration; for some bacteria and under
certain conditions it is permanent, i. e., it becomes a racial
characteristic and is transmitted by heredity from one generation to future generations of this bacterium; for other
bacteria it is less permanent, and after a time, they acquire
their former molecular structure.   Different classes of bacteria behave differently to attenuating agencies, from permanent changes on the one side to but slight changes on
the other, represent the extreme limits between which, at

various points, are found the individual limits of attenuation for various classes of bacteria. If, now, such stable substances as bacteria can be thus changed, it is certainly not unreasonable to suppose that such unstable substances as albuminoids can also be similarly affected, and that degree and duration of immunity will correspond to degree and duration of those changes produced in albuminoids by ptomaines.

The other phenomena of infection which remains for consideration—the varying degree of virulence which characterizes different epidemics—has also its analogy in fermentation, and its cause in attenuation. Experiment and observation teach us, that attenuation of ferment bacteria lessens their power of creating ferment products, and, that degree of attenuation, in this case, corresponds to loss of power in such bacteria. Analogy teaches that, in like manner, attenuation of pathogenic bacteria lessens their power of creating pathogenic products, (which are the true infectious agencies), and thus weakens their power of producing infection; but we are not compelled to rely wholly on analogy for a knowledge of this fact; the experimental work of Pasteur, of immortal fame, has furnished strong evidence in favor of this view. Attenuation of bacteria is induced by compelling these microorganisms to live and grow under artificial conditions that are unfavorable to their full development, and, it is often necessary to push these conditions to the extreme limit that is compatible with the life of the bacteria. Nature, also, has her methods of attenuating bacteria, which, like artificial methods, consists in surrounding the microbes by unfavorable conditions of life. These comprise sunlight, a pure and dry atmosphere, in fact, all these conditions embraced in the term climatic influences, many of which will vary with locality, population, climate, water and food supply, and sanitation. And, while all bacteria are more or less affected by these influences, some varieties of pathogenic bacteria are much more susceptible than others, for example, those which are indigeneous to the locality are, perhaps, less affected by their environment than are exotic bacteria which, like animals and other plants, thrive but poorly in fereign countries. Now, as malignancy of infection corresponds to virulence of the causative bacteria, and this again is modified by attenuation, it is seen how the environmental influences, which have been named, give to epidemics of infectious diseases their varying degree of malignancy, and, by the way, it also teaches the necessity for strict observance of sanitary measures, and how these may modify the virulence, or arrest the cause of the epidemic.

One more subject remains to be explained, and when this is done our task will be finished. We have presented the most prominent theories of immunity and have fairly and honestly discussed their claims. At the same time, these theories and the physical theory have been submitted to the same standard of comparisons that the value of each might be determined by its ability to philosophically explain the phenomena of infection and immunity, which are as follows, viz:

1. Phenomena of infection, (a) incubation; (b) self-limited duration and typical course of acute infections; (c) varying degree of malignancy of epidemics.

2. Phenomena of immunity, (a) natural immunity; (b) acquired immunity, (1) by single attacks, (2) by infectious bacteria, (3) by attenuated bacteria, (4) by bacterial products, (5) by blood serum and defensive proteids, (6) unclassified methods.

3. Inoculation fever.

4. Varying degree and duration of acquired immunity.
Answers were also required to the following, viz:

1. The cause for difference in failure of leucocytes, of
1st, susceptible, and 2nd, immune animals towards pathogenic bacteria; and

2. The nature and causes of biologic changes produced
in bacteria by attenuation.

The physical theory has explained all these subjects,
either in this paper or those which preceded it, with the single exception of that which relates to the incubative stage
of infectious diseases. These explanations are believed to
be complete in details and scope, to be uniform in methods,
and, furthermore, to be scientifically sound as they rest immediately on, or they are logical deductions from, accepted
principles of matter and motion. They are, further, sustained, in a remarkable manner, by sound and legitimate
analogies, therefore, if the atomic theory of chemistry, and
the undulatory theory of light can be accepted for the sole
reason that they most perfectly explain the phenomena of
chemistry and those of light, the physical theory should certainly not be discarded without reasonable cause.

A few words in explanation of the incubative stage of infectious diseases and we will have done. The phenomenon
of incubation, unlike the other phenomena of infection, has
no analogy in fermentation. The cause of this is no doubt
to be found in the conditions of the two processes, one of
which is relatively simple, while that of the other is quite
complex; e. g., a fermentable substance, say grape sugar of
vinous fermentation, is an organic, non-living substance
whose only bonds of union are its chemical and crystalline
forces, and when these are overcome by the waves of the
ferment we have the phenomena of fermentation. On the
other side, the albuminoids of the body are united, not only

by the force of chemical attraction, but also by those of organic life which must be overcome by pathogenic bacteria before infection of the organism can result. These forces comprise nerve and brain influences, food supply, and those other influences which give to the organism its condition of "good health," and, it follows as a corollary of this proposition, that influences which impair these qualities, such as fright, fatigue, starvation or unhealthful food, and depressing emotions, will diminish the resistance and thus open the way to infection of the body.

When, now, pathogenic bacteria are introduced into the body of a susceptible animal, we can understand why a period of time must elapse before infection takes place, for these bacteria have to increase in considerable numbers before they can exert the amount of energy requisite to overcome the normal resistance of the albuminoids.

To those who have kept abreast of the progress in medical science, and are familiar with the remarkable curative effects, in certain morbid conditions of the body, which have been achieved by Constantine Paul, Brown-Sequard and others, from hypodermic injections of normal tissue juices aseptically prepared, our physical theory will indicate a rational explanation of the modus by which these results was secured, and of a practice which promises to occupy an important place in future medicine. This subject, however, is not germane to that under discussion, and must, therefore, be passed by with this mere suggestion.

[THE END.]

www.ingramcontent.com/pod-product-compliance
Lightning Source LLC
Chambersburg PA
CBHW021528210326
41599CB00012B/1416